CONNECTION
GENERATION

How connection determines our
place in society and business

Iggy Pintado

Cover Design by Jason Cupitt
Author Photo by Ross Coffey

ISBN: 1-4392-2532-X
ISBN-13: 9781439225325

Visit www.amazon.com or www.connectiongenerationbook.com
to order additional copies.

What others have said about Iggy Pintado:

"Iggy has captured and explained in his book *Connection Generation* the impact of how we can participate in the new connection paradigm and the impact on our lives and business."

- *Gordon Makryllos, Vice President, Pacific and Managing Director, Australia & New Zealand, APC*

"Iggy is one of the world's first online networking experts who can talk from hands-on experience. He's changing the way businesses connect to their customers."

- *Malcolm Auld, Principal at Malcolm Auld Direct and one of Australia's most respected Direct Marketers.*

"As an experienced leader, coach and inspirational mentor to many Gen Y's, Iggy has the exceptional ability to harness the power of connectivity in online networking for the 'new world'."

- *Jennifer Taylor, Client Representative, IBM*

"Iggy is a leader in the development of the relationship between people, communications media and their interaction, for positive outcomes in networking and marketing."

- *Roger James, Chairman, Australian Marketing Institute*

"Iggy's knowledge is exactly what business owners need to get solid online strategies on how to grow their business in sales, customers and connections."

- *Jen Harwood, Business Champion*

"As a successful marketer and online expert, Iggy provides valuable insights on effective online networking - particularly useful for career management professionals."

- *Stan Relihan, Asia Pacific's Most Connected Headhunter*

ABOUT THE AUTHOR

Iggy Pintado is an internationally respected business leader, consultant, speaker, marketer, author and connection technology expert. Over the last twenty-two years, he has held professional, management, and executive positions in marketing, sales, operations, and online management at IBM and Telstra. He is co-founder of ConnectGen, a consulting firm specializing in online networking. As a "super-connected" networker on LinkedIn, Facebook, and Twitter, he has earned his place in the *Top 25* list of the most connected business networkers in Australia.

For more information, visit www.iggypintado.com.

This book is dedicated to all the generations of my family and friends for their love and support and for being living examples of the power of connection.

CONTENTS

PREFACE

The next big thing is here. It's not about rising and falling markets, globalization, and climate change—they're all factors of something much bigger. Something we've always had that will enable us to manage our ability to address these issues.

It's about connection.

Connections from the past, present, and future—across age, race, language, and belief dimensions—instinctively, instantly, and globally. Connections to each other, information, experiences, and ideas that will determine how we cope with the challenges and opportunities we face in society and business—now and in the future.

It's about a generation of people who lived in an era of great connectivity that I call the **Connection Generation**. It's about the role people and groups play and how attitude, behavior, and capability affect their place in society and in business. Throughout this book, I've used a SMART—*specific, measurable, achievable, relevant, and timely*—framework to best convey the idea of the **Connection Generation** to you. I'm *specific* about understanding the details of individual, group, and network roles. I use facts where appropriate to quantify the points I'm making. I define what needs to be *achieved* to address the challenges and opportunities in managing this generation. I tell real and *relevant* life stories—using both actual and composite characters and situations—so people can better relate the ideas to their own connection experiences. This book is *timely*; both in terms of uncertainty and prosperity, to better manage the challenges and opportunities that face both society and business.

Finally, the ideas discussed are referenced throughout the book. For easy review, I've included a Quick Reference section for the main concepts at the back of the book.

Welcome to the **Connection Generation.**

INTRODUCTION

On July 12, 2008, the world's oldest internet user at the time, Olive Riley, passed away at a nursing home in a suburb of Sydney, Australia—she was 108. She maintained a web log or blog, which she jokingly referred to as a "blob." She shared her daily musings and her rich life's experiences, which included raising three children as a single parent, surviving two world wars and the Depression, and working as a station cook in rural Queensland and as a Sydney barmaid. She posted more than seventy entries while boasting an international readership numbering thousands. In her final post, dated June 26, 2008, Olive noted that although she couldn't "shake off that bad cough," she had "... read a whole swag of e-mail messages and comments from my internet friends today and…was so pleased to hear from you. Thank you, one and all." She logged off at 108 years old.

Six days earlier, my father, Felix, turned seventy-eight years of age. He is of Spanish/Swiss heritage and was born in the Philippines. To maintain contact with friends and relatives in those countries, he set himself up with Instant Messenger (IM) software on his computer. The application allowed him to communicate to friends and relatives who were "online" at the same time by typing short messages to his family and friends at any time of day in real time on his computer keyboard. He was a proficient typist and an avid speller, having learned on a touch-type typewriter during his school days. It is exactly the same action of typing words on a keyboard— just like he used when he wrote letters to his family using his Olivetti typewriter many years ago. The difference is that this technology has an instant response and interactive component where family members type back on their computers over the internet as they receive the words—instantaneously. No waiting weeks and months for return letters and having to respond back the same way. He has since installed a web camera on his computer that allows him to see and speak to his family friends in real time over the internet as clearly as he can see people on his beloved television. He recently asked me what I knew about another emerging technology called Skype. He is seventy-eight years old.

In August 2009, I'll turn forty-eight—four days after United States President Barack Obama expects to reach the same milestone. Over the past thirteen years, I've searched for information, corresponded with friends, messaged family, shopped for goods and services, booked travel and entertainment, and made payments through my bank to energy, water, and telecommunications suppliers—all over the internet. Two years ago, I decided to travel to Asia with my family. I renewed my Australian passport in minutes by downloading a form on a web site. I also registered my physical whereabouts on the Australian government web site when I visited Bintan, a remote island in Indonesia, so that the authorities knew where I was in case something happened to us. In the past, I would've had to travel to a post office or call an administrative office to mail the necessary forms to me, reaching me in seven to ten days.

Almost a year ago, I reconnected with a person I used to work with while on a company assignment in Tokyo, Japan, in 1998 by using an online networking application called LinkedIn. Since then I've found many more past contacts and made some new ones along the way. At the time of publication, I had over four thousand direct connections on LinkedIn and over fourteen million. indirect connections. I also use the more social online networking site called Facebook, where I've connected to over seven hundred "friends." I was born in 1961, and the social behaviorists say that at my age I'm not supposed to use "new" technology as pervasively as I do.

I woke up one morning and found my eighteen-year-old son, Andrew—thirty years my junior—playing the popular online card game *Texas hold 'em poker* with people from Norway, Canada, and Spain. The night before, he completed a clean sweep of three online gamers from Singapore, South Africa, and Argentina while playing against them using online game FIFA 2009, a football/soccer game, on his Sony Playstation3. All the while, his mobile phone was by his side and at the ready in case he got a call or text message from his friends or his boss (in the event that he forgot that he was rostered to work). Most people expect this behavior from the average eighteen-year-old.

Due to the rapid adoption of these technologies, especially in the twentieth century, it is thought that generations—defined by the year in which one was born—determine certain attitudinal characteristics and behavioral tendencies toward society and business. Although there is no agreement among researchers as to the definitive generations of the last century, the most common generations are the pre-boomer (pre-1945), baby boomers (1946–1960), Generation X (1961–1981), and Generation Y (1982–2002). During my research for this book, I also found a traditional definition of the term generation is "the average interval of time between the birth of parents and the birth of their offspring," which makes a generation an average of around thirty years in length.

This prompted my thinking on the adoption of communication technology by generation and led me to ask how four people, all chronologically thirty years apart in age and from completely different generations, have a direct connection to people, places, and information using today's technology. To begin with, it is generally agreed that anyone between eighteen and thirty-five years of age is probably very proficient and comfortable with internet and mobile technology, having been brought up in the era of the personal computer and the beginnings of the internet revolution. It could also be argued that because this forty-eight-year-old spent most of his working life in the information technology industry he is by default "tech savvy" enough to understand and manage these modern technologies. Although the fact is, he is in the minority of people his age.

But should a seventy-eight-year-old really be expected to learn and master current information technologies and applications like instant messaging and web cameras, let alone spell IT? And surely anyone over one hundred years old has no place anywhere near a computer, let alone be proficient at blogging—an online activity a portion of the population still has no idea about. So what happened?

We're all connected. We always have been. We also know that some of us are more connected than others, which makes us realize that our degree of connectedness will have a profound impact on our social and business lives in an increasingly connected world. Whether you believe that a supreme being created life or that it was the result of a big bang that consequently led to our evolution from sea creatures to the people who roam the planet today, we all have a connection to each other. This connection manifests itself through physical attributes such as blood relations and places of origins. In John Guare's 1991 play *Six Degrees of Separation* one of the main characters, Ousa, muses to her daughter that:

"Everybody on this planet is separated by only six other people. Six degrees of separation. Between us and everybody else on this planet. The president of the United States. A gondolier in Venice... It's not just the big names. It's anyone. A native in a rain forest. A Tierra del Fuegan. An Eskimo. I am bound to everyone on this planet by a trail of six people. It's a profound thought... How every person is a new door opening up into other worlds."

Apart from the connections among billions of people in the world, we also have connections to information, objects, and concepts. As we trek through the journey of life, we gather items of value and experience that are the mementoes of our lives on earth. A prime example comes from the moment we start life. It is one of those common things we share as human beings. It's not just birth—but a "birth-day."

Each society has a range of rituals that signify these life events. Along the way, not only are the events and experiences remembered and recorded, but also objects and concepts are retained as symbols of our existence and subsequent connection. Most importantly, we know this because they're communicated through time via verbal or written reports and stories.

This has been the case for many generations across many cultures. The difference has been the diversity of retention of the connection and network records over time and their visibility and accessibility. For many cultures, there is a reliance on people's memories and the communication of the connection stories that link people, information, ideas, and experiences together. Over many years, humans developed tools and applications that facilitated the recording and storage of this information. First came painting, language, and the written word, which were mostly used to state and record facts but also to express and interpret events, experiences, and concepts. Subsequently, photography, the typewriter, the telephone, radio, film, and television enhanced media technologies and sparked a communication revolution like no other.

With each advance in communication technology, humans became more familiar with the need to use a variety of tools, devices, and applications to make them more effective. And as each technology evolved, the basic precepts behind the function of devices remained the same, although the complexity of their usage increased. For example, the first functional typewriter built by Remington in 1873 bore little resemblance to the more sophisticated IBM electronic version in 1973. Interestingly, the typewriter was the precursor to today's personal computer.

The world is now familiar with the *tipping point* concept introduced by Malcolm Gladwell in his book of the same name. He maintains that collective behavior, combined with factors such as connectors, stickiness, and context—at a given point in time—could combine to create a "social epidemic," which is commonly known as a trend. Connection technology tipped in the mid to late 1990s. Large masses adopted internet and mobile technology faster than any other previous technology. How? Firstly, we already knew about the power of connections—it was in our DNA. The structural importance of family, community, and company formed the basis of twentieth-century existence across societies, cultures, and businesses. Secondly, connectivity vehicles combined elements of previously tried-and-tested technologies such as print, mail, the telephone, the typewriter, radio, and television. This convergence allowed the rapid adoption of e-mail on a personal computer, mobile phone technology, the World Wide Web, and other media devices such as the iPod and PlayStation in very short spaces of time. Lastly, technology vendors cracked the code to provide access and search capability over the internet to what search engine maestros at Google called "people, places, and things." This combination revolutionized the way people connected to others and to information.

And it didn't matter when people were born or what generation—pre-boomer, boomer, Gen X, or Gen Y—they belonged to. If they were alive during this time, had direct or indirect access to a connection device and the relevant application—or someone who had one—they were connected. This is the *Connection Generation*. It's a generation that had previously experienced social and business connections primarily through face-to-face relationships. They had a familiarity with input and output devices such as telephones and computers to communicate. They now found a more cost-effective and time-efficient method through an instant, easy-to-access global connection technology platform to acquire and retain connections. It is the generation that caught the online bug and developed an appetite to stay connected ever since. This book profiles the *Connection Generation*—their evolution based on the convergence of communication and connection technologies. It explains their attitudinal, behavioral, and capability profiles both as individuals and in groups. It reveals a methodology to manage connected individuals via a PLAN and connected groups by being OPEN. It discusses the societal and business considerations and implications of our connectedness, and outlines some of the challenges and opportunities for managing, dealing with, and relating to the Connection Generation.

CHAPTER ONE
LIFE CONNECTION

Life isn't about finding yourself. Life is about creating yourself.
– George Bernard Shaw

1. It's a Small World after All...

It's quite common for people to respond with "It's a small world" when asked if they know someone or something that relates to some element of their personal backgrounds. But are we really that connected? In his book *Linked*, Albert Larabisi confirms our life connectedness. He notes that numerous network science studies denote links across many of life's forces including, "*... species in food webs appear to be on average two links away from each other; molecules in the cell are separated on average by three chemical reactions; scientists in different fields of science are separated by four to six authorship links, and neurons in the brain of the* C. elegans *worm are separated by fourteen degrees of synapses.*"

Larabisi also mentions another study conducted at the University of Notre Dame on the links among diverse web documents. The study reveals that "any document on the web is on average nineteen clicks away from any other." However, his most interesting observations are his references to social links. He references the beginnings of social connection research in a 1929 study by Karintay where the researcher observed that people are separated by at most five links. Larabisi also cites the theory of random networks defined in 1959 by Paul Erdos and Alfred Renyi who, in conducting a social interaction experiment of observed behavior among guests at a party, concluded, *"The guests are nodes and every encounter creates a social link between them."*

Karintay's work evolved into the more commonly known "six degrees of separation" in 1967 through the work of Stanley Milgram. Although he never used the term himself, Milgram's provocative question was whether one could measure the relative distance between strangers based on randomly selected individuals in the United States. He found the median number of connections between random contacts was 5.5, which was rounded up to six, hence the term six degrees of separation. The theory inspired a play, a film, a game, and the sixdegrees.org web site launched by actor Kevin Bacon.

The six degrees theory was validated in 2008 by Microsoft with research using their instant messaging (IM) application. A research team studied thirty billion instant messages sent by 240 million people in June 2006 and found that any two instant messenger users could be linked in 6.6 steps. "We've been able to put our finger on the social pulse of human connectivity—on a planetary scale—and we've confirmed that it's indeed a small world," Microsoft researcher Eric Horvitz said. "Over the next few decades, new kinds of computing applications, from smart networks to automated translation systems, will help make the world even smaller, with closer social connections and deeper understanding among people."

So how far back can we trace our connections? A 2007 study found that the source of the first people with blue eyes came from a small island near Italy. This group began a genetic connection that explains the origins of the connections among all human beings who have blue eyes. A 2008 *Time* magazine profile of U.S. President Barack Obama noted that his family roots could be traced back to both Republican President George Bush and Vice President Dick Cheney. My own family history reveals that I am a direct descendant of world-renowned fiction author Alexander Dumas. So where does connection begin at an individual level? A good starting point would be the connection to our birth.

2. Birth Connection

There is a physical connection in a human being's development between mother and child throughout the gestation and incubation period through the umbilical chord. Numerous studies show that a strong mental and emotional bond also develops during this time. When the baby is born, it formally connects with the world outside the womb by gasping for its first experience with oxygen, which it relies on to sustain life. It also experiences its first disconnection as the umbilical chord that connects it to its mother is cut. It's no surprise that the first reaction to the disconnection between mother ship and mother earth is one of bellowing cries.

In Western society, it is common practice to wrap the newborn baby in a blanket for the mother to hold close to her. The intent is to retain the feeling of security within the womb and to formalize the bond between mother and child in the external world. A name is given to the baby—sometimes visibly via a name tag placed around the baby's wrist or ankle while in the hospital—which will be its unique identifier from this day forward. In Western societies, the child's second name or surname is derived from the paternal side of the family to maintain a link to the "family." Once the child leaves the place of birth, it is registered and issued a birth certificate, which will be used as a proof of existence for this individual in terms of its place in society. Most societies have developed centralized registry systems that hold records of date and year of birth, parental identification, and place of birth details to assist in identifying, monitoring, and tracing all six billion inhabitants of the planet.

In most societies, each anniversary of one's birth date is remembered and celebrated as an annual event known as one's "birthday." Usually these birthdays are celebrated with family and/or friends. The celebrations are usually group participation rituals. These rituals have evolved over many years and differ depending on culture. What stays the same is that, for most people, they are remembered and recognized by their loved ones on the "special" day that they entered the planet. In Western society, the celebration involves food (birthday cake), drink (a toast to the celebrant), fire (blowing out candles), gifts (birthday present or greeting card), an event (party), and song (singing the birthday song). These celebrations are usually recorded for posterity using photographic equipment such as a still or video camera. It is common to have a more formal event when one reaches a milestone age such as a sweet sixteenth and coming-of-age events at eighteen, twenty-one, forty, fifty, seventy-five, and one hundred years of age. These events can be so memorable that some can vividly recall whom they were with, where they were, or what gift they received years after a particular birthday celebration. It is not uncommon to retain these memories by keeping photos and birthday cards as historical records of the event.

Due to the significance of the birthday connection that every human possesses, it also becomes a time for people who have not maintained contact to "reconnect." For example, it is my adopted practice to make a connection with my family, friends, and colleagues on their birthdays. This is facilitated by online networking sites such as Facebook and Plaxo that have reminder tools to prompt upcoming birthdays for your contacts, usually a few days before the event. I use this as an opportunity to keep in touch with my connections by calling them, sending a short note, or messaging them with a birthday greeting. It should be noted that said connections volunteer this information on their profiles so it is not perceived as "stalking them" once a year. Although I may not correspond with some of my connections regularly, at the very least, I initiate contact once a year on their special day.

As a child begins its journey into the world, there are certain elements that are integral to its development and connection to the world. Almost every new entrant to the world is born into a societal culture that is commonly linked to a place of family origin or religious belief. Eskimos, for example, who possess a very distinct culture and are indigenous to Alaska (USA), Canada, and Greenland, are believed to have a common ancestor in Asia with links to the Mongols of China and the Koreans. Regardless of this heritage, the average Eskimo child will grow up in the common cultural practice in one of a number of tribes that define Eskimo culture including Alaska's Inupiat, Canada's Inuit and Inuvialuit, and Greenland's Kalaallit. Despite this fact, the reality is that most of the world will define them as "Eskimos." The culture will have its own practices, rituals, and acceptable group behaviors, which the child is expected to adopt until it is old enough to make up its own mind as to whether to embrace the lifestyle for the rest of its life. The fact is that individuals are indelibly connected for life to the culture they are raised in.

The other individual connection to the world is nationality, or the country one either is born into or adopts. An Eskimo born in Alaska is linked to its native culture but is a resident of the United States of America. It is subject to the laws that govern that country regardless of tribal rulings due to Alaska being a state of the USA. If the child has a need to travel, it needs to apply for a passport document, which will be issued from the country of citizenship. Regardless of where individuals travel to in the world, they will be formally identified and recognized as citizens of a country and inextricably linked to preconceived ideas of that nation. Informally, people will form perceptions of them based on their external appearance.

Finally, there is spirituality. Almost every child born into the world is raised under some kind of spiritual thinking that it is exposed to in the early stages of development. In some cases, its spirituality defines its place in the world. For example, according to traditional Jewish law, a Jew is anyone born of a Jewish mother or converted to Judaism in accord with Jewish law. American Reform Judaism and British Liberal Judaism accept the child of one Jewish parent (father or mother) as Jewish if the parents raise the child with a Jewish identity. Jewish culture and its traditions are inextricably linked to the spiritual thinking and practice. The expectation is that people who define themselves as Jews participate in the various observances of the Jewish faith including prayers, holidays, pilgrimage festivals, and dietary laws. In the case of Jewish tradition, their spirituality sometimes tests the boundaries of a separate culture, nationality, and spiritual identity. For the rest of the world, all the perceptions connected to Jewish spirituality are assigned to people who practice it.

Another element that defines an individual is the social designation of a "generation," which is loosely determined by the year of one's birth. There is no agreement among sociologists on a precise definition of various generations, including generational cohorts. It is used by sociologists, researchers, and marketers to classify what the *American Heritage Dictionary* defines as "a group of contemporaneous individuals regarded as having common cultural or social characteristics and attitudes." Rightly or wrongly, it is used in society to analyze the attitudes and behavior of groups defined by the year of their birth, creating perceptions and labels for individuals belonging to each generation.

3. Relationships

Much has been written about the close ties among families, sometimes referred to as "blood connections." As time progresses, children move beyond their family circle and make friends at preschool and then school. These connections are random and sustainable relative to personality, social setting, and time. While most of these early connections are fleeting, some become "familiar" (derived from the word family) and can develop all the way through to adulthood. "Childhood sweethearts" grow up to share lifelong relationships.

For most of us, being sociable is a human characteristic. For almost all of us, it's a survival instinct to procreate. In every society, whether it is a primal urge or reason to raise children to maintain a succession line, there is a ritual that goes from dating to mating that occurs. In primitive times, partners chose each other based on purely physical attraction, hence the phrase "primal urge." Over the last few centuries and in many cultures, it was about ensuring one stayed connected to one's own class or group. In royal families specifically, there were strict rules as to whom you were able to marry. In the last forty years, there was still a public raising of eyebrows when Prince Charles, the heir to the throne of the United Kingdom chose Diana Spencer as his bride and subsequently, mother of the future king. There are arranged relationships as is the case in Indian culture. In some cases, the intended parties are predetermined to marry from as early as eight or nine years of age, meeting for the first time only days before they are due to potentially spend the rest of their lives together. In Western society, this prearrangement of relationships occurred in the era of what became known as societal "matchmakers," whose role was to match two individuals based on the class rules of the day. The term was popularized in modern times by George Bernard Shaw's *The Matchmaker* and by the popular Hollywood musical *Hello Dolly,* based on Shaw's play.

The most common dating process is the chance meeting at a social event or meeting place. The process involves both parties interacting through a self-introduction or an introduction from a mutual friend, an offer to buy a drink, or even a cheesy pick-up line. The intent is to find that connection that will result in future interactions, usually obtaining a contact phone number or e-mail address so that future connections can occur so they can get to "know each other." This process has the intended outcome of ascertaining whether there is a connection—mental, emotional, or physical—that may result in the act of mating for pleasure or procreation.

The process is significant in that there is much at stake. If one party shows any signs of indifference toward the other, it is seen as rejection. This can have devastating consequences if an assumed "connection" has been made by either party. The process to establish a "relationship" can take as short as a few weeks or up to years before the connection manifests itself. It goes without saying that the result of a successful match is—again either for pleasure or procreation—intercourse. The word has always been linked to its sexual context but has its origins in the course of interacting between individuals. The sexual act itself is a physical connection. I won't go into detail, as much has been written about the pleasure experienced by reaching a mutual connection point. The fact is that once the process ultimately results in mating behavior, a connection is not only the outcome but also the result. Today, a significant percentage of popular online web sites are dating sites, such as eHarmony and RSVP.com. Their entire reason for being is to mirror the "matchmaking" techniques of former times to facilitate the dating process and intended outcomes.

4. The Matrix

So how do all these life connection experiences allow us to make better sense of the world? I received an insightful perspective at a strategic marketing conference I attended in 1997. The event was also attended by senior marketing executives, including representation from the major airline, banking, and telecommunications organizations in the country. During the breaks in the formal sessions, attendees were encouraged to network and discuss the issues of the day, which is sometimes one of the more valuable aspects of these types of courses. At one particular break, one of the attendees raised the issue of the corporate matrix. He went on to relate how one of the fundamental roadblocks for conducting business within his organization was the fact that the organizational structure was designed to promote fiefdoms that were responsible for the success of their own particular product area or segment group, which, in many cases, collided with the company's overall vision, mission, and goals.

The reference to a matrix was that the organizational structure usually meant you had more than one boss, as the reporting and hierarchical lines were blurred by the siloed structure depending on a distinct focus. The specific example he raised within his company was around the establishment of an acquisition group whose sole purpose was to acquire new customers for the entire company. This group reported into a separate sales division, but horizontally to the other product and customer coverage segments, thereby having more than one accountable leader by organizational design and rendering that leader ineffective due to the multiple masters he or she had to serve. Other participants in the conversation soon chimed in, citing the exact same phenomenon was occurring in their organizations as a major impediment to conducting an effective and efficient sales and marketing function.

The next formal session was due to begin. We all sat there waiting for the speaker to arrive, and the audience began to stir as each minute ticked by that the speaker was late. Then from the back of the room appeared a man dressed in denim shirt and jeans, looking a lot like an academic who had never graced a customer's office. He bounded toward the lectern and, without apologizing for his lateness, began with:

"From the corporate research I've been conducting, the challenge of the organizational matrix in large corporations gets raised time and again. I just want to say up front that people who believe that it's unmanageable are...full of shit."

Predictably, the room fell silent, especially as some of us had discussed the very issue moments before. Here was a guy who didn't even introduce himself prior to arriving late for the session our companies were paying good money for, wearing the academic theorist's uniform, and starting the session by accusing us of being "full of shit" about the organizational matrix we had to work with on a day-to-day basis. After a brief pause he continued:

"We should all know how to manage a matrixed environment as we were born into one."

He then proceeded to deliver this memorable insight. He related how when most of us are born, we automatically have two managers—our parents. We learn from a very early age—sometimes instinctively—what we can get from each parent. In some cases, it's food, cuddles, and quality time from one boss and it might be playfulness and early morning attention from the other boss. But we know they are in charge—even when tested or challenged—as they provide the support and sustenance required for survival. As we grow and develop more experiences, we find out that they too have bosses, better known as the grandparents. They also provide support and love, but we seem to be able to obtain certain things from them that we wouldn't from our own bosses. Maybe it's because they are the managers of our own bosses.

Then there are other family players—brothers, sisters, uncles, aunts, nephews, and nieces. We learn through a mix of experiences and intuition what we can expect from each of them and sometimes what we can get away with. They are part of the support structure of our development and we learn to utilize them—or not—as a resource to achieve certain goals. Further development arises at school age where we start to make friends. We also experience our first "horizontally matrixed" manager in teachers and, spiritually, through priests, rabbis, or gurus. Once again, through observation and interaction, we get to know which ones will assist or just get in the way of progress. As we further progress through into our late teens, we encounter lecturers and tutors. Some even experience their first real business manager as they start a casual job, whether it is delivering newspapers, picking fruit, or acting as a gopher in a local office. We also get to experience life as a customer, from both sides of the fence. Consistently, we are learning and adapting to personality, leadership, and management styles that may or may not suit our needs.

The speaker then concluded that by the time one reaches the world of business, one has already received the best real life—and real-time—training to operate in a matrix environment. Over time, you've been prepared to innately manage a wide range of stakeholders, including vertical and horizontal executive and senior management; colleagues, peers and partners; and potential prospects and customers. The speaker ended the session by introducing himself as an organizational behavior expert. His name is not relevant, as his message was the key outcome. The message was that we go through our early development connections in a random yet structured form that manifests itself in how we prepare ourselves for the survival techniques required to deal with society and business.

5. Creating Yourself

There's a quote at the beginning of this chapter by Irish playwright George Bernard Shaw that says:

"Life isn't about finding yourself. Life is about creating yourself."

It's become a personal favorite as an appropriate catchphrase for what I define as the *Connection Generation*. During the 1970s, there emerged a "subgeneration" that became known as the "me" generation. Individuals defined their place in the world from a single, introspective, and "selfish" view, developed from an "out-looking-in" perspective that remained largely invisible to most and where self-interest was their "be-all and end all." It spawned a self-help and discovery industry that introduced yoga, meditation, and a range of books and courses on how to "better oneself." This may have been the beginnings of individuals wanting to find meaning in their lives and subsequently finding their place in it.

Just before the end of the last century, a generation evolved that was less about self-interest and more about creating both a real AND a virtual world persona. It was more about a plural, multi-dimensional, external, and "self-less" view. The "self-less" view stemmed from connected experiences, opinions, achievements, and emotions that reflected an "in-looking-out" perspective. Most importantly, these social perspectives were recorded on connection technologies known as online networks that gave it selective but nonetheless public visibility. What was previously an "invisible" view of their connectedness to the world was transformed to a "visible" one, opening up individual opportunities to better participate, collaborate, and contribute to the greater society and business world as never before.

The capability for individuals and groups to manage through their personal development and to interact as groups and communities evolved from the communication revolution that occurred in the best part of the twentieth century and manifested itself in the last five years of the century, giving life to the *Connection Evolution*.

CHAPTER TWO
CONNECTION EVOLUTION

Today, after more than a century of electric technology, we have extended our central nervous system itself in a global embrace, abolishing both space and time as far as our planet is concerned.
— Marshall McLuhan, 1964

1. Communication Revolution

Communication began when people first began to express themselves, either one-on-one or in small groups. They interpreted their lives in their own ways through the resources they had available to them. The first tribes of cave dwellers painted images on the walls of their dwellings providing visual interpretations of their hunting expeditions. Through the years, the development of language helped this expression evolve into the use of words and grammar. But there was a need to more accurately record the stories, thoughts, and ideas so they could be shared and passed down to the generations. After a few millennia, the next advance in communication addressed this need in the form of writing. Writing provided the ability to record information, thoughts, and ideas, extending communication to the education of current and future generations and beyond mortality. Communicating could be personal through a letter or diary entry or more formal through documents and contracts for business. Despite these advances, communication remained one-way or at best, one-to-few.

With the invention of the first printing press in 1439, Johannes Gutenberg instantly changed the communication dynamic and revolutionized the ability to share information beyond a single speaker to small groups. With the ability to empower thoughts, democratize education, and spread ideas beyond regional boundaries, printing became the catalyst for modern-day mass media. The written document—personal or public—could now be shared across borders. The *Diary of Anne Frank*, published in 1947, remains one of the most potent works of modern literature that would have remained a personal reflection had her plight not been shared as a story of the times she lived in.

With the sharing of ideas, people now had more to talk about in communities beyond communal word-of-mouth and idle gossip. In America, the town hall meeting became a forum where communities could share their thoughts and ideas, and debate and discuss the issues of the day. The limitation of communicating beyond regional borders was stretched when, in 1875, after much experimental work on voice transmission over wires, Alexander Graham Bell's United States patent for "Transmitters and Receivers for Electric Telegraphs" was granted, ushering in the era of the telephone. The emergence of the printing press through to the telephone, were important stages in the revolution in communication that moved from one-way to two-way and one-to-many messaging.

In 1896, Gugliemo Marconi was granted a British patent for the next advance in communication, radio. Its early application included Morse code used between ships and land and to pass on orders and communiqués between armies and navies during the First World War. Broadcasting became feasible in the 1920s, with the first commercially available radio receivers introduced in Europe and the United States. In 1938, American Orson Welles caused the American public to panic during a radio broadcast of a narrative rendition of the H. G. Wells classic, *War of the Worlds*. The power of this mass medium to connect with audiences was further realized by the likes of Adolf Hitler, Winston Churchill, and Mahatma Gandhi, who utilized public radio broadcasts to influence, persuade, and incite their constituents.

The motion picture materialized during this time and, with "talking pictures" in 1930, became a dominant entertainment vehicle. After the Second World War, the movie medium triggered a more distributable technology known as television, which, as an extension of radio, was designed primarily to broadcast sounds and pictures back into one's dwelling, as pictures did for the caveman eons earlier. This phase of the communication revolution—telephone, radio, and television—progressed two-way communication via the more effective medium of sending content via wires and electronics and customized communication based on perceived mass need through broadcasting.

2. Everything Old Is New Again

During the 1960s and 1970s, computer researchers were exploring the ability to share data between established information networks. This led to the development of "inter-networking," which became known as the internet. Following its introduction and commercialization in the 1980s, the internet grew in popularity throughout the 1990s with the rise of electronic communication by e-mail, text-based electronic catalogues called web pages, and the links between these documents through what became known as the World Wide Web. In the early 1990s, with the exception of e-mail, the internet was a large repository for information, which, once accessed, could be viewed by the user. It was the largest collection of information on the planet. At this same time, mobile phone technology became more pervasive.

By the 1990s, almost all previous communication technologies had a more efficient and effective digital equivalent. Expressive communication forms such as photographs were virtually replaced over time. The traditional photo—a memento of the past—was becoming a relic. The picture camera, with the capability to store a maximum of thirty-six images on plastic film, which in turn needed to be processed—usually at a photo lab that took more than twenty-four hours to process—no longer worked for most time-sensitive consumers. The advent of digital photography with the ability to take up to seven times as many pictures, store them electronically, and selectively print them on a digital printer in moments at the users' discretion, led to the demise of the Polaroid camera and its cohorts in the printed images era.

Hand and typewritten letters that were sent and delivered to far-off locations became affectionately known as "snail mail." They were replaced by the more efficient fax and then e-mail technology, which allowed a user to send correspondence electronically, have it delivered in seconds, and potentially obtain a response shortly after, at the recipient's discretion. Users warmed to e-mail as a messaging application, as it used a keyboard interface similar to the familiar typewriter—in the form of a QWERTY keyboard—which represented the first six letters on the second row of a traditional keyboard. The telegram had suffered the same fate earlier, replaced by instant messaging (IM) on a computer that, as the name suggests, allowed short personal messages to be sent in real time electronically to an online recipient.

This capability was replicated by an even more effective application known as the Short Messaging System (SMS), which became a standard application on a mobile phone device. The telephone still had its place, ironically through the dependence of some users on a landline to access broadband technology required to connect to the internet at home. The telephone continues to be threatened by such technologies as mobile phone and Voice over Internet Protocol (VOIP) that is increasingly making the humble telephone less of a necessity. In fact, the International Telecommunications Union predicted that there will be four billion mobile phone users by the end of 2008.

Another technology holding its own is radio, although it has evolved over time. The content of audio broadcasts continues to be threatened by the adoption of portable digital devices such as iPods (personal music) and digital radio (online access to worldwide music stations). Once again, the irony of radio remaining a relevant technology to the masses is the fact that its most popular format is interactive, or two-way. Talk radio, also known as "talkback" radio, which encourages listeners to call, e-mail, or text message their opinions and thoughts to the program, remains the radio show format that attracts the widest radio audiences, as popularized by American political commentator Rush Limbaugh and comedians Howard Stern of America and Australia's Hamish & Andy. Even popular television was being given a run for its money by interactive cable channels. Once again, it's ironic that one of the most popular TV formats in the last few years has been "reality TV," which is dependent on viewers choosing to vote for and against the dismissal of cast members through an interactive process that involves registering your vote using text messaging or "voting" on an online web site.

Toward the end of the century, the communication revolution had served its purpose. It taught people to begin to express themselves using various tools and to share information and ideas through various applications and provoke their thinking through broadcast media and the internet. The world now knew how to meet their communication needs in one-way, two-way, and mass mediums. They now wanted to match this ability with their innate and intuitive need to connect to each other. The compelling social and business need to optimize the world's richest resources—the collection of people, information, experiences, and ideas—by connecting and interacting with it was ready to be tapped. In fact, it was ready to tip.

3. Tipping Point 95

Malcolm Gladwell describes a tipping point as "the moment of critical mass, the threshold, the boiling point." If you had to pick a date when communication technology began to tip, it would have to be October 13, 1994. This was the date that the gateway to the internet—the Netscape Navigator Web browser—was launched publicly. It effectively opened the door to the wealth of information being amassed on the World Wide Web. In 1995, it caught the imagination of the online world with users employing the browser to access the rich information that was posted on this developing web of knowledge.

But there was a need for more than just access—people wanted the capability to search the whole base of information to find what they were looking for. In 1998, a company named Google provided this exact functionality. If Netscape provided the door to the World Wide Web library, Google swung it open, took users across the threshold, and took them on a guided tour of every level—page by page, book by book, and resource by resource. For the first time, the world could not only access information, they could search using specific words and phrases for what they were looking for.

This ability to interact with information instantly—using any browser—through an easy-to-use interface (the single input text box that is Google's front page) and for free, was the catalyst for the world's connection to, as Google itself put it, "people, places, and things." It enabled a generation of users to navigate and explore the web universe of information. Netscape and Google were no longer for the geeks and nerds who had made it their domain. It made people change their attitude and behavior to being online—forever. Businesses began to sniff out the opportunities. Sites such as eBay and Amazon.com pushed the envelope by pioneering e-commerce, allowing users to buy and sell goods and services over the internet through their online channels. IBM expanded online business to the entire corporate world by naming it "e-business"—and branding it with their own distinctive red "e."

As the world experienced the internet's triumphs and tribulations through its infamous boom and bust period in the early twenty-first century, another global event tested the new technology's ability to connect the world. When terrorists attacked mainland United States on September 11, 2001, it jolted the world into the reality that communities weren't as connected to people, information and ideals as first thought. It made people realize that there were elements in society that didn't care to share in the freedom that others enjoyed—in fact, they wanted to limit and destroy it. People then started to search for perspectives and answers. They wanted to know the truth about these events and they realized that this information was not solely available from a single country's media agencies but from multiple sources located around the world. The events of 9/11 put the internet into context.

Post 9/11, two other applications emerged that would continue to contextualize people's global realities via connection. In 2004, Jimmy Wales launched Wikipedia, a web site that collected written information on anything and everything through user submissions globally. It became the people's information resource catalogue—written by users, for users. In October 2008, there were over 2.6 million articles written in English and many others in a variety of diverse languages and dialects. In 2005, YouTube did the same for the video medium, with users posting their collection of personal and publicly available video for the entire world to experience. By July 2006, more than one hundred million videos were being viewed every day with two and a half billion videos watched in June 2006. In January 2008 alone, nearly seventy-nine million users had made over three billion video views with approximately ten hours of video uploaded every second. According to a *Time* magazine article, the phrase "video snacking," the practice of watching online videos during breaks in the day on a computer or mobile device, entered the lexicon.

With other connection technologies being introduced and commercialized during this time, an umbrella term to define the sharing of user-generated content through technology became known as social media. It arrived through various applications including blogs, vlogs, podcasts, wikis, online social and business networking, virtual reality, and microblogging. In each case, regardless of text, graphics, audio, or video formats, users could interact with any posting by commenting on the content. Users were even able to mix and match some of the information and applications and present them on a single site via technology called mash-ups. The capability to interact with the technology also became known as Web 2.0.

Within the space of fifteen years, the web evolved from a singular communication to an interactive medium. The static web page that transcribed hard copy content onto a web page with minimal to zero capability to respond to the data was the design point for Web 1.0. Web 2.0 promoted two or more ways to interact with people and data. The ability to communicate back to other users and systems by searching, sharing, creating, collaborating, messaging, networking, blogging, poking, voting, etc. made online activity all about interacting with the people, messages, and ideas presented. These mediums transformed communication as we knew it previously to what I call connection technology. These technologies provided more than just the traditional capability to send and receive messages. It gave individuals the power to create, choose, manage, and control the what, when, how, and why of connecting to people and information—and they took this power with both hands, a keyboard, and a mouse.

4. *Now Media*

For many years, *Time* magazine ran their Person of the Year award. Each year, the writers would choose the most impactful, notorious, or powerful individual for the past year. Previous winners included historically significant figures such as Albert Einstein, Mahatma Gandhi, Pope John Paul II, Mikhail Gorbachev and Nelson Mandela. In 2006, *Time* magazine's Person of the Year edition featured an image on the cover of a computer with a reflective material on the screen that resembled a mirror. Its by-line announced, "Yes, you. You control the Information Age. Welcome to your world." *Time* magazine's *YOU* cover, although controversial, reflected the view that the power of communication had shifted to the user, as facilitated by connection technology. The average individual could now choose what, when, and how to connect to people and information, instantly and ubiquitously.

Online media is referred to as "new" media. A more accurate description would be *NOW* media. We wanted to be part of the action, we wanted to connect, and we wanted it now. We didn't want to be limited to taking thirty-six pictures on our cameras. We wanted to take as many as we wanted so if we made a mistake we had a selection to choose from. And we didn't just want to take the picture and have someone else process it. We wanted to be able to see it, share it, print it, and store it so we could find it, anytime. We wanted to interact with it. Good photos were no longer subject to photo spoilers such as the "red-eye effect" and unwanted people lurking in the background. We could now fix these problems with easy-to-use photo editing software—in an instant—that was able to specifically correct red-eye and crop people and items out of perfectly good photos.

We didn't necessarily want to wait for the news to be delivered by daily newspaper that was printed the day before or even wait for video of a global event. We wanted access on our terms so we set up news feeds and forced agencies to update sites as events happened. This included streaming video of important events that we could view immediately after it occurred so we didn't have to wait for the evening news broadcast to see for ourselves what happened. We didn't necessarily have the time to travel to events and conferences, although we were very interested in the content. We wanted the content to be available on our terms—when it was convenient, timely, and relevant. We were looking for solutions like webinars, podcasts, and even telepresence, all of which allowed us to be there virtually if we couldn't be there physically. We wanted to share what we had and knew on our own terms. It became a way to define our place in the world. We wanted to send photos, stories, and videos to our family, friends, and colleagues as soon as we created them or received them. Some of us even wanted to share them with the whole world through blogs and sites like YouTube.

Our need to connect extended to social and business relationships. We wanted to renew old friendships and relationships and selectively stay and keep in touch. We wanted to manage these connections on our own terms. We chose to flock in droves to MySpace, Facebook, LinkedIn, and a wide variety of online networks, as we found these platforms useful in providing this capability.

Most recently, we wanted all this to be handy. We didn't want to wait to get home or to the office to send an e-mail, look up a contact, view a map, or search the internet. We wanted it to be as available as our keys, wallets, or purses. We wanted it all accessible on a mobile device. Telecommunications companies like 3 in Australia were advertising Nokia's latest internet-enabled phone with the headline "E-mail the boss or Facebook the goss." That's why in 2008 Apple sold a million of their latest model iPhones within three days of their release.

5. Power of Connections

We always had an innate desire to connect. The communication revolution taught us how to speak to each other individually or en masse using various mediums. Connection technology allowed us to make it easier, more effective, and efficient to connect. It provided the capability for people to interact conveniently, instantly, and pervasively. If people chose to, they could reach people and information how they wanted, when they wanted, and anywhere they wanted. It made connecting SMART—*Specific* to our areas of need, interest, or expertise; *Measurable* to the point of knowing how many connections or contacts, post views, and votes on the content; *Achievable* in terms of realizable goals and outcomes of what we wanted to get done; *Relevant* to what we wanted—not necessarily what someone else wanted us to know; and *Timely* so we could get it instantly or when we wanted to receive it.

All this power and capability would have a discernable im-pact for both society and business and gave rise to a new generation. In the years following 1995, anyone who lived on this planet with access to a communication or connection technology device - regardless of age or ability - attained automatic membership of the *Connection Generation*. At the very least, they possessed a basic connection profile into the online world and, at best, clearly visible and definable con-nections to one or more people, information sources, and ideas. They were more connected to the world's population and information resources than those who lived at any pre-vious time. And they didn't take it for granted.

CHAPTER THREE
CONNECTION IS THE APPLICATION

As of now, computer networks are still in their infancy...but as they grow up and become more sophisticated, we will probably see the spread of computer utilities, which, like present electric and telephone utilities, will service individual homes and offices across the country.
— Leonard Kleinrock, UCLA Press Release, July 3, 1969

I. Exploding Web

The internet as a connection technology caught on fast. To achieve a market penetration rate of 25 percent in the United States, telephones took thirty-five years. It took television around twenty-six years, personal computers almost fifteen years, and mobile phones about thirteen years. It's no surprise that internet adoption achieved that rate in just seven years. It became a worldwide phenomenon. According to the Worldwide Internet User statistics updated June 30, 2008, 76 percent of the world's internet users resided in twenty countries. The countries with the most internet users were China, United States, Japan, India, and Germany. The countries with the highest percentage of users per head of population were Canada, Australia, Japan, United States, and South Korea.

Why were they accessing the web? One reason was that it literally spoke their language. In 2007, Google set itself a target to make more of its products available in multiple languages, starting with the forty languages read by over 98 percent of internet users. By July 2008, they had developed thirty products in thirty languages including multiple local dialect versions. In the same month, the company launched Google Aotearoa, the language of New Zealand's indigenous people. People were interacting online via their desktop and laptop computers and their BlackBerry and iPhone mobile devices. By the middle of 2008, car makers in the United States and Europe were developing new models with unrestricted internet access for release within the year. People weren't just logging on and browsing content as they had been in the past—they were actively using it.

In a presentation titled "*Predicting the next 5,000 days of the Web,*" delivered at the TED conference in 2007, Kevin Kelly reported that the web generated one hundred billion clicks per day, eight terrabytes per second of traffic, two million e-mails transmitted per second, one million instant messages per second—and 5 percent of the world's electricity consumption. We were all using it regardless of age or sex. According to a May 2008 study conducted by BlogHer and Compass Partners, 35 percent of women eighteen to seventy-five years of age in the United States participated in reading, writing, or commenting on blogs at least once a week. Studies reveal that in Australia, the country with the second-highest internet penetration rate based on population size growth, 70 percent of older people had used the internet at least once.

In the early part of the twenty-first century, people's lives had become busier with less time to get things done. We needed to find the most efficient and time-effective ways to do what needed doing. We figured that if we could connect tasks with people and information, it would take less time. Once the technology became available to do this and we knew how to use it, we would use it. Connectivity became what IT companies defined as a "killer application" for this generation. So what were we connecting about?

2. Transaction Time

The smart operators knew that with all this connection activity, there was money to be made on the web. Google and Yahoo used the traditional advertising model—selling strategically placed advertising on commercial web sites on their search results that users could click through—to bankroll their product developments. Google made its first billion dollars in less than six years compared to McDonald's, which took twenty-four years. It must be noted that there were victims of the 2000 crash, such as fashion venture Boo. com that lost $160 million before being liquidated and the ill-fated Pets.com that raised $82.5 million before closing months later. But rising from the ashes of the crash were some of the old stayers like Amazon.com and eBay, which quenched the thirst of the online faithful by using the web to create a virtual marketplace where vendors could reach a global audience for items that would be advertised in local papers, garage sales, or specialist retail stores—and where consumers could refine their searches for products and buy them at special "web-only" prices.

In June 2008, The *Sun-Herald* newspaper in Australia conducted a survey on a number of online products that were also available at large retail stores. The report revealed that "…in some cases, products were more than twice as expensive in local department stores as they were on overseas web sites." There was still a lingering concern regarding the security of transacting online with dodgy operators looking to catch out unsuspecting users. But the old rule of "buyer beware" was just as relevant in the e-commerce world as it was in traditional trade.

A range of buyers from the humble bargain hunter to the affluent wanted to transact online, albeit for different reasons. A 2008 Google study found that online shopping was the preferred retail channel for the rich. The study surveyed the shopping habits of over a thousand people with incomes or net worth greater than a million dollars. Of those surveyed, 91 percent said they always or often looked at online reviews before buying luxury goods and spent over $114,000 per year online compared to $23,000 spent in retail. Around 94 percent agreed that "making a high-end or luxury brand available online doesn't cheapen their opinion of the product or brand." It seemed that users had a preference to complete simple transactions online. This included registering their cars and boats, applying for personal, home, or car insurance, booking their travel arrangements, trading shares on the stock market, and even making medical appointments. Web operators offered specific services that somehow looked more appealing than their traditional equivalents.

Sites like worldbaggage.com.au and personalporter.com.au offered travelers door-to-door luggage delivery with a guarantee of no luggage lost at around half the price of what airlines charged for excess baggage. You could arrange to have the bags arrive at the hotel beforehand and, in the case of personal porter.com.au, you could sign-up for text message updates to your mobile phone so you could track the bags along the way. Busy people wanted an alternative to contacting suppliers by phone and inevitably having to wait in a phone queue for their queries to be addressed. The web provided vendors with the capability to publish frequently asked questions (FAQs) with most sites providing e-mail contact addresses—both of which were accessible twenty-four hours a day, seven days a week. In 2008, Alaskan Airlines launched a virtual travel assistant on alaskaair.com where, by typing in your questions, virtual receptionist Jenn responded in written text and spoken answers, as well as asking her own clarifying questions.

3. Find and Be Found

The web had the power to connect people to information—any information—in a practical way. If you wanted to find the definition of a word, you could visit dictionary.com or many other online versions of traditional dictionaries like Webster, Oxford, and Macquarie. You could also just type in the word "define" on any Google search box plus the word you were looking for, to be presented with a list of possible definitions from various general and specific sources. If you weren't familiar with some of the language of the internet, there were web sites like commoncraft.com that provided plain English explanations of the terminology, using short videos. If you weren't technically inclined and needed someone to repair or maintain your computing devices, sites like geeks2u.com.au in Australia could have someone come to your home or office and fix it on-site. If you were looking for that out-of-print book, sites like Faber.co.uk in the United Kingdom could reissue titles from its backlist and, using print-on-demand technology, reprint orders for a single book.

The web was a place to find and be found. A friend of mine, Joe Talcott, told me that he was creating a web site to post the numerous family letters, photos, newspaper clippings, and other mementos he had accumulated over time. It was a labor of love as he tried to make sense of the order of the material so it expressed the fullness of the family story. He told me about an interesting connection he made while doing some research on the web. Joe's late father used to tell him the story of a calf that he was entrusted to raise when he was in his early twenties and living in Illinois in the United States in 1939. The calf—named King Bessie Senator—became a prized All-American bull from 1939 to 1948 and never lost a competition. He was a grand champion and later sold for $8,000, which was big money in those times.

Joe thought that there had to be some information on the web about grand champions of the past. His brief search not only revealed a site that traced the calf's family tree, its name was listed on a trivia site and he found a photo of the bull's headstone on Flickr.com, a popular online photo management and sharing web site. He sent a message to the person who posted the photo who then shared other photos of King Bessie and the farm where he lived. The web had provided Joe with a set of links to the past that included his family, their interests, and their history.

Ancestry sites are popular places for users to trace their past. A British site—ancestry.co.uk—compiled the details of the 8.9 million settlers and migrants who were sent to Australia and were available for search. You could find the details of your family tree and history in a short space of time. In the case of King Bessie, even find pictures of dead bulls from the late 1930s on this World Wide Web. With the ability to search and find information, the web also assisted with real-life connections. In June 2008, a story in the Australian *Sun-Herald* newspaper told of the cat that was found after being reported missing since 2005. The microchip on the cat revealed the owner's name, but an attempt to contact the owner found that her telephone had been disconnected. The authorities then tried Facebook, found the owner's page, sent her the message that her cat had been found, and the pair were reunited in days. In Britain, a couple who met on an online dating site and were planning to marry, found that they had lived seven houses apart on the same street for seventeen years.

4. Comment, Compare, Collaborate

The web promoted and encouraged mass participation—and people were prepared to share their perspectives. Google cofounder Sergey Brin related on his personal blog the discovery that his mother was diagnosed with Parkinson's disease and by testing himself found that he was also susceptible to the disease. He shared that he was compelled to "*... have the opportunity to perform and support research into this disease long before it may affect me.*" By 2008, anyone with an interest in sharing news, information, experiences, or ideas was publishing, reading and/or commenting on a blog. In the single month of August 2008, a comScore Media Metrix report found that blogs had 77 million unique visitors in the United States, compared with 75.1 million MySpace and 41 million Facebook visitors. Like some comments made in life, they led to arguments. So sites like sidetaker.com provided the opportunity for users to state their positions on topics using virtual discussion boards and let thousands of readers take or oppose their views, by registering their yea or nay votes. As the arguments stayed virtual, any physical expressions of anger remained offline.

As people continued to voice their opinions online, they also wanted their perspectives to be counted and compared. This gave rise to online opinion polls and surveys, including Australia's pureprofile.com, which allowed users to create a profile and record their preferences for various products and services. They were paid cash to complete relevant surveys—and were made offers by relevant vendors—for items matching their preferences. At ratemylife.net, visitors rate themselves and their lives by answering questions that would assign scores in the areas of accomplishments, virtues, sins, and mental and physical states. At ratemyprofessor.com, you could rate your current and past college professors on attributes such as helpfulness and clarity. By 2008, over six million students had rated a million professors across the United States, Canada, England, Scotland, and Wales.

And if you really wanted to know what you were worth in cold hard cash terms, humansforsale.com could give you an estimate after you completed a test that *"will attempt to place a value on your life using a variety of criteria in four basic facets of life...(including) athletic ability, education level, income, amount of exercise, weight, and sense of humor."* It provided a warning of its subjectivity and scientific inaccurateness. Comparisons turned into online competitions. Games manufacturers developed the next generation of computer games so gamers could play each other online and on mobile devices. In 2008, Spore, a life-simulating game from the makers of the popular computer game The Sims, allowed you to create a creature and guide it through its five stages of evolution—cell, creature, tribal, civilization, and space—while interacting with other user-created creatures along the way. If you were a karaoke fan, the PlayStation3 version of Sing-Star provided the capability to buy and download the latest songs from the online SingStore and to upload and share videos of users' performances.

The web became the place to personally collaborate—to meet, share, and keep in touch. Social networks made it easy to do so. In 2007, my niece from Madrid and her friend from Barcelona in Spain visited Australia and traveled from Perth on the west coast to Sydney in the east. On their flight across from Perth, they chatted with a guy they were seated with for the duration of the four-hour flight. Upon landing in Sydney and while waiting for their luggage at the baggage carousel, they asked me to take a photo of them with their "new friend" before he departed. After saying their good-byes, the guy casually said, "*See you on Facebook.*" While it sounded like a throwaway line to the uninitiated, the fact was that they would connect through the online networking site and they regularly message each other to this day. Here were total strangers, meeting briefly on a flight in a foreign country, now separated by thousands of miles yet socially connected on a single application on the World Wide Web.

The web as a meeting place thrived through online dating sites including America's eHarmony and Match.com and Europe's Meetic. Specialized sites like SawYouAtSinai.com would "*combine Jewish matchmaking with online Jewish Dating to bring you the ONLY personalized, private and online service for Jewish Singles.*" In Japan, Tokyo company Ai Senior (translated as Love Senior) represented three hundred clients fifty and older who were members of a six-thousand-person online database of single and available elders.

People even reconnected in death. I worked with Michael at IBM in the late 1980s but hadn't seen him or stayed in touch over the past ten years. He met his partner, a lovely woman and fellow IBMer named Beth, at a business function. They eventually married, settled in the United States, and started a family. In 1998, while on business in New York, I had lunch with Beth and Michael who introduced me to their beautiful newborn daughter, Jayde, who I held in my arms. Ten years later, I received an e-mail from Michael telling of Beth's passing after a seven-month battle with breast cancer. The news stunned me to my core and I felt helpless to express condolences to Michael and Jayde being so far away in my home base of Sydney, Australia.

In the same e-mail, Michael also mentioned that he had set up a commemorative web site for Beth—bethellenby.com. The site allowed people to light virtual candles, post personal condolences to the family, and share memories of Beth in written form and through photos. Despite the tyranny of distance and time, one could still share in Beth's memory as part of the grieving process while celebrating the life of a loved one. The site remains a sad memory for me but a fitting—and everlasting—tribute from her family and the many friends of her life. I thank Michael and Jayde for the opportunity to reconnect me with Beth, even after she had physically disconnected from the world.

5. Greatest HITS

In the early days of the web, a dominant measure of site effectiveness was the number of "hits" a particular web page would get from visiting users. It's ironic that the behavioral profile of the connected individual is best summarized via the acronym *HITS—hunt, interact, test/trial, and share.* The web changed the dynamic of how people connected—socially or in business—to other people and information. They applied connection technology so much so that their attitude, behavior, and capability in an online environment changed. They had the capacity and the choice to connect like never before and they transformed into connected individuals. Connected individuals who acted and thought differently than the "Webutantes" in the early days of the online world. Their search and research missions of the past became hunting trips. They wanted to not just search, seek, and explore, they wanted to *hunt* down the specific person or piece of information they were after.

The rise of numerous global and specialized dating sites—a social activity that once relied on chance meetings—became a more disciplined filter for prospective partnerships through connection technology. It didn't matter if you were Jewish or Japanese, you could search and find that elusive "perfect partner" online. Once users found their "treasure," they wanted more. They wanted to *interact* and engage it by finding out more about the subject through various vehicles including face-to-face meetings, written, pictorial, audio, and video formats. They sometimes proceeded to present, publish, or post the information in their own way, as Joe Talcott did with King Bessie. If they weren't sure, they were prepared to *test* and *trial* it like the early pioneers of the commercial web, where the winners won big and the losers lost even bigger. They would use simulators like Second Life, Spore, and The Sims to replicate life virtually to try out experiences they wouldn't dare to in real life.

Finally, they were prepared to selectively *share* the spoils of their finds. If they found something they thought others would value, they'd invariably tell someone, somewhere by sharing it online. Michael valiantly shared the memories and stories of his beloved and immortal Beth as did Sergey Brin about his potential life-threatening legacy. The proliferation of current-day blogs numbering more than seventy million per month in 2008 is online proof that people were prepared to share their experiences, thoughts, and ideas with the masses.

This *HITS* profile explained the behavior of the connected individual. But surely we were all different to a degree in terms of how we hunted, interacted, tested, trialed, and shared. And it was—we discovered that we all had different connection profiles that determined why we behaved the way we did.

CHAPTER FOUR
CONNECTION PROFILING

I can never be what I ought to be until you are what you ought to be. This is the interrelated structure of reality.
– Martin Luther King Jr.

1. Three Rs of Connection

In the years after 1995, there was a distinctive behavior change in the way people connected to others and to information. On the web, they enhanced their searching to hunting down their desired informational prey. They weren't satisfied with just finding data of interest; they wanted to interact with it. They were less afraid to try out new technologies and applications leading to more downloads of trials like anti-virus applications and smaller applications called applets. Once they acquired this new knowledge, they didn't hesitate to share it, selectively or to the masses. People got the *HITS*. This change in behavior begs the question, what caused it? Why do people want to connect and be connected?

There are three basic factors at play in determining a person's attitude and behavior toward connecting to people, information, experiences, and ideas, which I call the *Three Rs of Connection: Reason, Relationship, and Return.* The first is *reason*—put simply, why people connect. People usually possess a compelling need or want to develop a connection. The objective for searching Google is usually to find information out of interest or curiosity. The rationale for posting a profile on an online network is for the specific outcome of sharing information with others. Other reasons for social or business connectivity may be a perceived inherent value in getting connected.

Motivational expert Tony Robbins speaks of the six basic human needs of which the fourth is love or connection, which he describes as the need to feel either for someone or a group. The desired outcome may be for a single purpose or a combination of reasons. Some common reasons for personal connection are proactive courtesy, which is defined as keeping in touch with friends, relatives, and colleagues on a regular basis via the use of an online social network; diplomatic etiquette, which is exemplified by the unspoken obligatory communication practice while traveling of sending family, friends, or colleagues postcards from places being visited, even while on a relaxing holiday; and a business requirement, initiatives related to retaining customers, with the rationale being that the better one is connected to a customer, the less likelihood of that customer switching to a competitor. As rational beings, there are many conscious or subconscious purposes that exist for connecting to people and information, which is defined as *reason.*

The second R is *relationship,* which defines one's connection relative to circumstance or place in the world. An example is the long-distance relationship. In the years preceding the pervasiveness of telephone communication, relationships between people who lived many miles from each other but wanted to conduct a relationship was sustained primarily by written correspondence. People who wanted to be friends only became known as "pen pals" while the more serious relationships produced documents that were referred to as "love letters." As telephone communication became more pervasive, geographically challenged relationships migrated to communication via voice calls, which became known and marketed as "long distance calls." Although this technology was more expensive for the average user than the humble letter, the personal element of listening to and responding to someone's voice coupled with the "instantness" of response was a key adoption factor.

Fast forward to today's connection technology where a combination of these mediums is present in online connection. E-mail is the electronic child of written letters and documents. Instant messaging and text messages are direct descendants of the telegram. The long distance phone call via a landline is slowly being replaced by Voice over Internet Protocol (VOIP) technology, a connection application like Skype, which allows both audio and video interaction via a computer network. In business, Cisco's telepresence technology combines audio teleconferences with state-of-the-art video technology that gives the impression that remotely located meeting members are in the same room. CNN used this technology during the 2008 presidential election day coverage to connect reporters and guests from various remote locations. In both these cases, relating via simulating a face-to-face conversation was the key reason to connect. Although technology may become more sophisticated and pervasive, the compelling motive to connect remains the maintenance of a social or business *relationship*.

The third R of connection is *return*. Again, put simply, it's what a person or group gets out of connecting. For online networkers, a common return is the recognition by the community as someone who is "connected" either socially or in business by the number of "visible" connections they have. In the offline world and from a social perspective, it's more about who these people hang out with or are seen with. The social pages on most tabloids are full of party photos of the perceived socially connected. Those who make regular appearances are often referred to as being on the "A list," which assumes that their celebrity or fame gives them a "must invite" tick on the major social event invitation lists. Those categorized on the B or C lists include socially connected has-beens, pretenders, and persona non gratas.

Today, through online networking sites such as MySpace, Facebook, LinkedIn, Plaxo, and Xing, the number of friends or connections is clearly visible to people one is connected with. LinkedIn, in particular, has a subculture element of connecting to be recognized, evidenced by the existence of groups such as TopLinked.com, which regularly publishes the users with the most connections and encourages users to climb the ranks of "most connected" through a paid subscription model. The reason to connect should deliver a tangible benefit or return to the connected individual.

2. Attitude/Behavior/Context

So people connect for a reason, based on their relationship motives and with an expectation of some kind of return. But how do we know if these connections are effective in their dealings and experiences in society and business? Here's where the *ABCs* of connection, namely *Attitude, Behavior, and Context,* kick in. In his book *The Tipping Point* Malcolm Gladwell argues that social epidemics were usually started by influential personalities, which he defines as the "Law of the Few." He proceeds to discuss a particular profile he calls "connectors." These are people who are socially connected to their communities, action oriented, and have the ability to successfully transmit important communications effectively and efficiently.

As an example, he relates the story of the legendary American Revolutionary Paul Revere, famous for his midnight journey warning fellow Bostonians that the British troops were amassing for an attack and that they needed to be prepared. His famous midnight ride became a catalyst for raising awareness of the British army threat and succeeded in preparing and building the resistance that was instrumental in "winning" the larger battle that became the American Revolution. It's argued that Revere's attitude and behavior toward "connecting" to his constituents in not only delivering the message but creating a platform for action through his passion and enthusiasm raised the level of importance of his message to the community audience that received it.

Gladwell then contrasts Revere's achievement with the lesser-known William Dawes. While Revere was responsible for spreading the word over the west side, Dawes was given the task of warning the constituents of the imminent British invasion on the east side of the affected area. Gladwell notes that Dawes was "less successful" due to his "unconnectedness" to the community he was delivering the message to. It was argued that the critical components of Dawes's attitude and behavior, combined with the context of the situation, played an important part toward his task and the conclusion that his mission was not as effective as Revere's. In the cases of Revere and Dawes, the point was that there are "degrees of connectedness" based on factors such as attitude, behavior, and especially the context of the situation. The question is: what are these degrees of connectedness and what are the practical impacts on society and business of having one profile over another? I decided to conduct a short study to assess whether one's perception of the word connection was related to attitude and behavior toward being connected, which I define as connection profiles.

3. Connection Perception

Why are some people perceived to be or actually more connected than others? Is there a connection gene that some of us have—and that others don't—that makes us more or less connected in various degrees? Could it be based on one's personality profile or attitude toward being connected? Or is it a chosen or adopted pattern of behavior that makes someone connected?

I conducted a social experiment to attempt to measure one's self-perception of what connection means and compared it to people's actual and published connectedness. I surmised that there was a link between attitude and behavior toward connection that made me want to test the theory. In September 2008, I posed a simple question to a predefined group of colleagues, friends, and family, eighteen to eighty years of age and from different regions globally. The question posed was: *When you hear the word "connect" or "connection," what image or picture comes to mind?* I asked respondents for a "first-thing-that-comes-to-mind" answer. I collected responses via a combination of face-to-face interviews and e-mail responses. I received 130 survey responses and analyzed the data.

I found that the responses could be grouped into five broad areas. There was the CLASSICAL image typified by responses such as nature, love, the universe, the Vitruvian Man, and Michelangelo's *Creation* painting, which depicts the hand of God reaching out via a single finger to the finger of man. The second category was CONCEPTUAL imagery, with responses such as puzzles, connecting dots, and building blocks. The third category was RELATIONAL depictions, which incorporated responses such as hands touching or handshakes, people kissing, and lightning bolts or sparks. The fourth category was PHYSICAL, with responses such as chains, links, switches, wires, and even sex. Finally, STRUCTURAL images appeared in the form of networks, process maps, family trees, DNA structures, and organization charts. A small percentage of respondents chose a mix; however, the first suggestion was taken as the first answer.

I began by profiling the responses by age and grouped them into the most common generational groups being: pre-boomer (pre-1945), baby boomers (1946–1960), Generation X (1961–1981), and Generation Y (1982–2002), of which I took approximately equal groups of thirty members in each category. I then looked at each respondent's "visible connectedness" to people by using the number of published connections on online networks in which I had access to their connection profiles. I used an arbitrary scale ranging from an individual's 0–50 visible connections, 50–100, 100–150, 150–300, and 300 or more, and ranked each respondent into these five groups of "connectedness." Each of the visible connectedness groups had a proportionate mix of generational profiles, with the exception of the 300 or more connections group that contained no pre-boomer generation members.

Going into the experiment, my hypothesis was that people from specific "generations" would gravitate toward certain images in line with their generational profiles. For example, the pre-boomer and baby boomer generation would pick a more classical interpretation of connection imagery, such as Michelangelo's *Creation*. Similarly, I expected Generation X to have a more conceptual or relational perspective due to their claimed logic and reasoning prowess. As Gen Y were born and raised in the hyper-technology era, I assumed that they would possess a more physical and structural outlook.

What I found with this sample group was that the analysis was not definitive in grouping generations to specific images, with little correlation to the generation they belonged to and the image chosen. The pre-boomer generation group's highest image score of CLASSICAL was 40 percent, the baby boomers scored a little better with 33 percent favoring CONCEPTUAL imagery, the Gen Xers could only manage 24 percent for RELATIONAL and 27 percent of Gen Y chose PHYSICAL definitions.

The key finding of this particular experiment was that there was a significant correlation between the connectedness groups and their first response to image selection. The data showed that people with 0–50 connections where likely to choose a CLASSICAL connection image (75 percent), the 50–100 connections group chose CONCEPTUAL images (80 percent), the 100–150 connections chose RELATIONAL pictures (88 percent), the 150–300 group chose PHYSICAL (84 percent), and those with 300 or more connections chose STRUCTURAL images (90 percent). The study revealed in a very basic way that the hypothesis of one's self-perception of connections coupled with actual connectedness could potentially be grouped into various attitudinal and behavioral segments regardless of generational bias. I began to define these as connection profiles. These profiles were less about generational characteristics and more about adopting, choosing, or gravitating toward a profile based on a mix of personality, attitude, and behavior toward connectivity. Other factors—and the combination thereof—including desire, passion, discipline, and context also played roles in remaining or progressing to another connection profile.

4. Profile Factors

I did some further modeling on the findings into the connection profiles. I started with the visible connectedness groups and matched them with their frame of reference vis-a-vis their responses to mental pictures of the word connection. I then applied a series of categories including base needs analysis, patterns of observed behavior, attitude toward technology, time spent on connection-related activities, and the connection applications or devices that best exemplified profiles. I assessed how these factors could be grouped and matched to people I knew, as well as to famous personalities who are perceived to be connected, so as to provide a common frame of reference to describe the various profiles. As I had the data available to me on the connectedness groups due to my visibility of their public online network contacts, it was practical to use this filter as a determining factor to define a profile. The base assumption was that if people took the time to work on the public visibility of their connections through an online networking tool, then that would be an indicator of their connectedness behavior.

I once read one of those social riddles popularized by publications like *Reader's Digest* that posed the question: *Why are computer aficionados and drug addicts both commonly referred to as users?* I can only assume that the element of need plays an important role in determining their attitudes and behaviors to how they "use" or apply their specialization, whether, in the case of the riddle, it was a personal computer or heroin.

As part of the experiment, I decided to conduct some follow-up interviews with selected respondents in each connection profile to ascertain what their base needs were in becoming connected. Consistently, the major needs could be classified into five basic groups. The logic behind using their perceptions of connection through images or pictures provided an indication of people's attitudes toward connection. The assumption is that if their definitions of connection were classical or conceptual they would have a less hands-on approach to connectivity. Similarly, if their images of connection were more physical or structural, their propensity to become more directly connected would be evidenced by their connectedness behaviors. The patterns of observed behavior filters were based on the respondents I knew well, who were mainly family, good friends, and colleagues I had worked closely with. In some cases, I conducted selective clarifying interviews to ascertain the behavioral characteristics associated with connection. I used publicly available data in researching the famous people.

Once I had the observed behavior information, I applied a sociological model known as the technology adoption cycle to assist in determining attitudes toward technology. This model describes the adoption or acceptance of a new innovation according to psychological characteristics of defined adopter groups. Although there have been various interpretations and a range of adaptations to this model, I've assumed a slight variant of the original model developed by Bohlen and Beal in 1957. The model assumes that the first people to actively use a new technology are classified as "innovators," followed by the "early adopters" and then the "early majority." The next group is the "late majority" followed by the "laggards." I then proposed a view of time spent on connectivity by using a mixture of theoretical and actual data. It was implied that if people had actively worked on converting a large number of colleagues, friends, and family to visible connections on an online network, then they would have spent a significant amount of time on sending invitations in building their social or business networks. Similarly, if people had a small number of visible connections or none, then it could be construed that they spent a minimal amount of time on connection activities.

I conducted a search to ascertain whether specific research had been conducted in this area. Colmar Brunton, Australia's largest independent research agency, is a leader in developing online research communities to enable clients to gain fresh insights and accelerated innovation. A 2008 study of 1,500 individuals conducted by Colmar Brunton explored the relationship between level of online connectivity and demographic/psychographic profile. The study showed that individuals with 150+ connections were more likely to be Gen X or Gen Y. However, the study revealed that people who perceived themselves to be "not connected" had high ownership rates of mobile phones (94 percent), laptop or desktop computers (93 percent), and internet access (87 percent). It can be concluded that although the "unconnected" may not have visible connections to people via an online network, they possessed connection technologies that allowed them to connect to people and information.

It must be noted that the power of context, introduced by Malcolm Gladwell in the *Tipping Point,* is a significant factor in determining an individual's connection profile. Gladwell made the point that social epidemics *"...are sensitive to the conditions and circumstances of the times and places in which they occur...and can be reversed, can be tipped, by tinkering with the smallest details of the immediate environment."* I believe the same to be true for the attitude and behavior of a connected individual. If the context of an individual's circumstance changes—sometimes even in the slightest way—the person is subject to a profile shift based on the given situation.

5. The Five Connection Profiles

My research revealed that there were five distinctive profiles that defined an individual's degree of connectedness. These groups included *basic, passive, selective, active,* and *super* connectors. The *basic* connector has the least number of visible connections, has a CLASSICAL mental image of connection, is a laggard when it comes to technology adoption, and usually delegates connection needs to a known person. The *passive* connector has a moderate number of connections, has a CONCEPTUAL view of connection, is in the late majority when it comes to technology, and exudes "passenger" behavior when it comes to connecting. The *selective* connector has a manageable number of connections, has a RELATIONAL view of connectivity, is in the early majority of technology adoption and is engaged in connection behavior. The *active* connector has a large number of connections, has a PHYSICAL mental picture of connection, is in the early adopter category of technology, and drives his or her connection activity. As the title suggests, the *super* connector is hyper connected, has a STRUCTURAL view of connectivity, is an innovator with technology use, and possesses a disciplined management style toward connectedness.

The following chapters describe each of these profiles in more detail using examples relative to the *three Rs of connection*. It includes a brief description of their attitudes, behaviors, and contextual dimensions. Each profile also contains a brief discussion of their capabilities and the implications of managing and dealing with the group from social and business perspectives.

CHAPTER FIVE
BASIC CONNECTORS

A hidden connection is stronger than an obvious one.
– Heraclitus of Ephesus (Greek philosopher, 580-40 BC)

Geoff is a taxi driver in his early fifties who has been driving taxis around Sydney, Australia, for the better part of his life. I use his taxi services to transport me from my home to the airport and back, both for corporate and holiday travel. I've known Geoff for over fifteen years and during our many conversations over the years, we have discussed what he knows about and how he behaves in the online world. He admits that he is a "dinosaur" when it comes to technology proficiency and very much behind the times of personal computers and the internet. He admits that he doesn't understand the technology, as he is baffled by the "lingo" and it seems to be a younger person's domain. His preferred technology device is the standard two-way radio common in most taxi cabs. For many years until the present day, this has allowed him to receive broadcasts of available jobs around the area he happened to be in, bid for the particular jobs he was able to get to, and stay connected to the taxi driver community in an "invisible" way.

Mary-Jane is a teacher who just turned forty and works at a school in a country town in the state of New South Wales, Australia. Her attitude to communicating and connecting is limited by the resources available to her in this remote town. In her town, the post office and telephone are the base devices of connection. The mobile phone is becoming more pervasive and the broadband connection is due to be upgraded from very slow to slow. She has an e-mail address that she looks at from time to time if she remembers and is prompted by someone who has sent her something relevant and important. Her preferred technology device is a telephone, which she uses to stay in touch with her family and friends in the city.

Damian is a plumber and, at twenty-one, he has just finished his trade apprenticeship. While his friends talk about the cool things on the internet like online networking sites such as Facebook, instant video sites such as YouTube, and the awesome interactive games you can play, he doesn't have an online presence and does not participate. He has a standard mobile phone that he uses for both work and his social life, giving out his number to only the select few he chooses to do business with or his close social circle. He uses the mobile to communicate with his own friends and family, usually to discuss face-to-face liaisons, like drinks with his mates. He prefers text messaging, as it allows him to "speak" to people one person at a time, in his own time, and relatively cheaply.

* * *

Geoff, Mary-Jane, and Damian are *basic* connectors. Despite being from various generations, *basics* feel more comfortable with tried and tested ways of communicating and connecting. They possess a very traditional view of the word "connection," which manifests itself in their frame of reference being the more CLASSICAL interpretation of the word. They usually see the connection with images of nature and creation. The attitude of this group to handling connections is based on the adage of *"if it ain't broke, don't fix it."* Their perspective is that the established way is much easier and more practical than anything others are used to and they have no desire to try out newer and potentially more efficient connection technology—if they don't have to. It's easier for this group to excuse themselves from taking the time to learn a new method of doing things due to their comfort level with the ways they are familiar with.

Although the majority of this segment can be classified in the "older generation" of people in the pre-boomer and baby boomer generations, there is a fair representation from Gen X and even Gen Y, depending on their preferences and needs. Although familiar with modern devices, *basics* struggle to fully comprehend the application of devices to their social and/or business situation. They have a strong preference for face-to-face or voice call interaction and usually correspond by writing personal postcards, letters and—if they know how—text messages. The devices that personify this group are the two-way radio and the old walkie-talkie. On two-way radio, once you are on a desired channel frequency, it's relatively easy to operate. Once activated, you listen to one person speak and then press a button to speak to that person. Similarly, the walkie-talkie is in the same category of usage but portable. It should be noted that these devices are primarily single application only—that is, calling/receiving only and, interestingly, are the precursors for today's standard and more pervasive mobile devices. If these individuals own a mobile device, it is probably an older model with the most basic features—in fact, similar to the ancient walkie-talkie.

This group usually has a "visible" core of zero to fifty connections. The visibility of these connections is usually found in more traditional contact tools such as a written address book or business card file. There may be more visibility of these contacts if they happen to use applications such as e-mail or mobile phone contact lists—which in the majority of cases are set up for them by a family friend or acquaintance. The irony is that this group may have many invisible contacts stored in contact address records such as the classic black address book, business card file, Rolodex (remember them?), or even an e-mail contact list. However, these are largely unmanaged and usually outdated, as there is little to no discipline to keep them updated with current information. They possess a good social group with whom they may regularly interact, but they manage their "connections" in a random and ad hoc fashion. Their process of staying connected to people is based largely on memory prompts, based on comments such as *"I haven't heard from Henry in a long time, maybe I should give him a call"* and the occasional written diary entries such as "Playing cards at Daisy's place next Thursday." The most famous *basic* is U.S. Senator and Republican presidential candidate John McCain. In August 2008, McCain admitted that he was technologically illiterate, having his staff print off his e-mails for him, and that he relied on his wife, Cindy, to make sense of the computer for him.

* * *

Their attitude toward technology is as a "*laggard,*" which is often described as conservative and, by definition, the last people out of the proverbial connection technology blocks. They claim to be technologically disabled and are usually in deep denial about their ability to learn and use technology to connect. They are reluctant to understand the workings of the technology, usually put attempts to explain it in the "too hard" basket and during conversation may mentally "turn off" easily. Their preference is to spend as little of their time using connection technology as is humanly possible. The basic connector is also challenged with applying technology to process. As technology is not at the forefront of their thinking and they are entrenched in "established-way-is-the-best-way" thinking, they may not see the possibilities and benefits of using technology to save time, money, or effort.

Taxi driver Geoff related this typical basic connector behavior story to me with regard to applying a process using a connection technology. Due to the busy nature of his job, Geoff forgot that the deadline to pay his boat and trailer license had crept up on him and he now only had twenty-four hours to renew the license to keep it current. He did not want the license to lapse and was in a panic the day before it was due as to how he was going to get the time to renew it. He spent time contemplating various scenarios in his head as to how he would juggle his day job with the need to get to the boat registration office, join the inevitably long payments queue once he got there, make the renewal payments, and then get back on the job.

He got home that evening and remembered that his twenty-something son would be home and could potentially run the errand for him. Later that evening, he related his license renewal deadline dilemma to his son. He asked his son whether he could assist by managing the *travel to-queue-then pay* process at the registry office on his behalf. Without wasting another moment, his son asked for the boat registration papers plus the credit card he planned to make the payment with. The son disappeared into his room and Geoff thought, "*Great, he's agreed to do it for me.*" Moments later, his son reappeared and handed his father a printout that confirmed the payment and said that the completed boat license papers would be sent in the mail. Within ten minutes, his son completed the entire transaction that Geoff thought with travel, queuing, and payment process time included, would take him the better part of two hours to do.

The basic connector is a *delegator*. With multiple offline connections, the basic connector knows at least one person in the other connector profile segments. This is usually someone who is online savvy whom the basic relies on as a "connection agent" or "technology translator" to assist in managing and interacting in the online world, on the basic's behalf. Due to the basic's perceived inability to use and apply technology to interact online, the basic proactively finds ways of delegating requirements to a contact who has the "online knowledge."

The best example is the tech-savvy grandchild who briefs and educates the grandparent on using anything from search on Google to enrolling in online social networks. We have a family friend in her forties who has never familiarized herself with connection technology. She shares an interest in knitting with my wife. One day, she "found out" about a web site called eBay, where she was told she could find and buy reasonably priced knitting materials from around the world. Instead of learning how to access, search, and interact with the popular auction site, she requested that my wife do so on her behalf to obtain the materials she wanted.

Mary-Jane the teacher uses her students to do her more advanced "Googling" for course information. Damian the plumber asks his girlfriend, who happens to be a Facebook user, to find out about local parties that have been organized via this medium, on his behalf using his girlfriend's online persona. Geoff realizes from experience that his son is an *active* connector (discussed later in the book) and now "utilizes" this connection to delegate some of his day-to-day tasks.

It should be noted that when *basics* take the time to learn and become more familiar with technology that will assist them with their day-to-day processes, make them more effective and efficient, and that they deem of value, they have a greater willingness to adopt it. An example is a review of taxi driver Geoff's business operation. What intrigued me about Geoff's setup was the number of additional devices he had recently been using to run his business, for a person who was a self-confessed technology "luddite." His beloved standard two-way radio was still installed. However, there was also a personal digital assistant (PDA) device that allowed him to receive and respond to e-mail bookings from regular clients and a mobile electronic payment device that let him process online credit card transactions and print receipts. Geoff also had a wireless listening device temporarily attached to his ear, which would not look out of place in any science fiction movie. This device allowed him to answer calls from his mobile phone without interacting with the handset while he was driving the taxi or away from his vehicle during a break. "My Bluetooth," he proudly informed me.

The proliferation of devices in Geoff's taxi is a case of once the value of the technology is explained, adoption follows. Once he was sold on and most importantly trained on how to use the application of the devices, he not only adopted but embraced them. The PDA device facilitated the customer interaction with his business by providing clients the option of booking his services via e-mail, allowing Geoff to access those e-mails and respond to the customer in real time. The mobile payments device let Geoff process customer credit card transactions on the spot and saved him the late evening administration process time of reconciling hard copy receipts with completed jobs. The wireless headset device created a hands-free connection between his cellular phone and the physical mobility he required. Once he was familiar with the devices, they became easy to use.

Another example is Helen, my dear mother-in-law. Helen's a pre-boomer, enjoying a life of retirement, with a self-determined aversion to technology. She does not own a computer and has no intention of learning how to use one. She does not own a credit card and refuses to use an automatic teller machine (ATM). Despite this, she possesses a convection oven that requires a pilot's license to operate, a knitting machine that has a built-in computer that she navigates with the skill of your average computer geek, and regularly uses her mobile phone when she's out and about. Each of these activities is specific to task—cooking, knitting, phoning—and she will deny that she has any more than the most rudimentary knowledge of how to operate these devices to complete tasks. This includes the minimum amount of interaction time with said devices.

Basics will invariably own a mobile phone but their "usage" of said device varies depending on absolute necessity. While Geoff is reliant on using this medium from a business perspective due to the nature of his trade, most *basics* will most likely stay with an older/ancient mobile phone model with only the most basic of features such as making voice calls (dial number using number keys then press green button) and accepting calls (press green button went it rings). They will not necessarily use voice mail or text messaging capability, as this would be "too hard to manage."

In many cases, *basics* become frustrated with additional functionality that they cannot use. The other frustration is with the need to recharge their mobile devices when the battery is low—even if they "haven't used the phone for a while"—as previous technology such as their telephone did not require them to take this action. This frustration may lead to forgetting to turn the device on, not carrying it with them, or even abandoning the use of mobile phones altogether leading to frustration from friends, family, and colleagues.

Basics use a single device to apply to the connection process. Teacher Mary-Jane uses her telephone exclusively for voice calling her friends, family, and colleagues in other areas. She could utilize the same phone line to connect either a fax machine or an internet-enabled computer to send photos, write letters, and share documents. Her choice is to use the phone for voice only, as that's what she's comfortable with. The same applies to retiree Helen with her techno convection oven to cook only. The look on her face when one explains that ovens of the future will be internet connected, which will let you start the cooking process, regulate temperature settings, and turn the oven off from a remote control device like her mobile phone, from anywhere such as when she's out shopping, is nothing short of priceless.

At the same time, as more applications become available on mobile devices and this group has more familiarity with mobile phones, the mobile platform may be the catalyst to progress basic connectors to the next level. In the June 2008 edition of Australian *Reader's Digest*, a contributor tells the story of his seventy-six-year-old father who, despite still having good driving skills, had trouble judging distances to the point of "driving by sound," which he defined as not realizing an object was there until it made the sound of being hit by his car. His son invested in a reverse-sensor car kit, which would send out an audible alarm signal when the car was getting too close to an object. So he installed the kit for his dad and sat in the car with him to test it for the first time. As his dad backed out of the garage, the alarm tone went off—and his dad reached for his mobile phone. Single device equals single application.

* * *

Context and relativity are important factors for *basic* connectors. The context of their connection is relative to their singularly focused need to connect to people, information, and ideas. For example, the context of a person teaching in a remote town with little exposure to technology, let alone infrastructure, may mean that the person has to exist with the limited physical social connections in the community. However, the corporate businessperson who relies on relating with customers and prospects in a tangible way as well as interacting with technology in the fashion that business stakeholders are familiar with, may not survive if his or her attitude and behavior toward connectedness is at a basic level.

From a social perspective, the inability to keep pace with technological advances and perceptions of being considered a social outcast are the biggest challenge for this profile. Despite their richness of offline connections, with limited visibility of connection to individuals or groups online, the *basic* individual could be perceived as a connection outcast. Tradesperson Damian may choose not to engage with his social group online but as having an online persona becomes more prevalent amongst his friends, disengaging may mean being left out of important social information, conversations, and events. Similarly, from a business perspective, the *basic* worker needs to stay relevant to retain a place in a connected world. Increasingly, it will be a matter of survival in a connected business world as more customers, prospects, partners, suppliers, and employees gain increased knowledge and expertise in being visibly connected. Taxi driver Geoff acknowledged this trend in his adoption of a PDA device that allowed his customers to e-mail him their bookings. He realized that if that's how his customers wanted to connect with him, he needed to adapt or miss the opportunity.

The proliferation of applications that promote connectivity, collaboration, and interaction, coupled with the availability of easier-to-use and more mobile hardware devices, will make it increasingly more difficult for individuals to remain in the *basic* segment. The situational and value context of applying connection technology to process encourages behavior that is modifiable, if the individual is willing to learn the value of adoption. Just as taxi driver Geoff was able to apply technology to a business process, he is capable of learning how to adopt connection applications to his other business and life processes. Once the technology infrastructure becomes available to teacher Mary-Jane, she will have more options to stay better connected to the stakeholders of her life.

Once *basics* have the confidence, attitude, and discipline to manage their connections, they can migrate to other connection profiles. However, they must balance their traditional ways and their delegation behavior with their adoption of applicable technology to their particular circumstances. Conversely, in some cases, the technology can get the better of this group due to their "not-wanting-to-learn" characteristic, which, in some cases, means they stay a *basic* for an extended period of time. In an increasingly connected world, this may not be a sustainable profile depending on lifestyle and career choices.

* * *

SUMMARY: Basic Connector

* Attitude—traditional, conservative, laggard, delegates
* Perception—classical
* Contacts—some to multiple, but usually invisible
* Connections—0–50 visible, but knows at least one connection agent
* Connection technology—1–4 devices, but selective for application only
* Direct time spent—minimal
* Examples—taxi driver Geoff, teacher Mary-Jane, plumber Damian, Helen the mother-in-law, and Senator John McCain

CHAPTER SIX
PASSIVE CONNECTORS

At that time I was using a mouse pad from the Apple Library in Cupertino, California, famous for inventing and appropriating pithy sayings and printing them on sportswear and mouse pads... The one I had pictured a surfer on a big wave. "Information Surfer" it said. "Eureka," I said, and had my metaphor.
— Jean Armour Polly

* * *

Diana is a Fortune 500 company senior vice president in her mid-fifties. She is responsible for all direct and indirect sales, marketing, and service provision for her company's product offerings to existing customers and prospects in the small and medium business market. When I worked in her division, I was a regular participant conducting reviews of the group's online sales and service strategy, for which she was also accountable. At one of the reviews, she volunteered her view of how she managed her online presence. *"Although I understand what it is and why people would, I don't really have a need to use it regularly,"* she said.

She revealed that as an ultra-busy, time-poor senior executive, she spent a lot of time on the phone to customers and managed her e-mails via her trusty real-world personal assistant, Seamus. She once used Google to search for a company profile she was researching but was spooked by the 280,000 search results that came up. Seamus was much more proficient at locating information for her quicker than she ever could. At home, her partner managed all their online banking and bill payments. She didn't have time to establish a presence on an online social network despite her children spending hours on end on MySpace and Facebook. *"Once,"* she admitted, *"I did use the internet to order some groceries but I gave up as it gave me too many options to choose from and took too long."* Her preferred technology is her standard mobile phone, which she uses to make and screen her calls when she chooses to. She only receives text messages and rarely, if ever, sends one out. She was starting to use a Black-Berry so she could access her e-mail remotely but needed convincing to use it regularly, as her assistant screened her mail for her anyway.

Loretta is in her mid-forties and has management responsibility for organizational development at a medium-sized organization that services the Asia-Pacific region. Her attitude toward communicating can best be described as reactive, responding only when she feels a compelling need to. She uses her mobile phone as a device to make outgoing calls and prefers not to answer her phone, letting the caller leave a message on voice mail. She returns calls only if she absolutely has to. She has an online networking presence on both

LinkedIn, for business, and Facebook, for social purposes. She admits that she joined, as she was invited to by a friend or colleague and *"wanted to see what all the fuss was about."* She has a minimal amount of connections, visits the sites only when there is some information she's been prompted to access, and her participation is limited to accepting invitations from people she knows, rarely inviting people herself.

Sam is an accounting student currently completing his university degree. At twenty-one, he knows the importance of having an online presence on a social network. His girlfriend coerced him to join Facebook by convincing him that it was an efficient way to connect to family, friends, and social information such as party invitations and gossip. He is very picky about who he connects with, limiting connections to family and friends. Sam doesn't prioritize his social participation on the network. He's given his girlfriend access to his Facebook account and relies on her to "clean up" his profile, accept invitations, and monitor the social environment on his behalf. He will participate in some activities such as Facebook armies but gets freaked out when his girlfriend's father pokes him on the odd occasion.

Diana, Loretta, and Sam are *passive* connectors. Once again, despite having diverse generational profiles, *passives* have an understanding of communication and connection applications but choose not to prioritize them in practice as part of their day-to-day regimens. They possess a very theoretical view of the word "connection," manifested in their frame of reference as being a more CONCEPTUAL interpretation. They visualize connection as a puzzle, "connect the dots" or converging arrows images. They manage connections with a "lean back" attitude, which means they prefer to watch, listen, and absorb what is being presented to them before connecting with the person, thought, or idea. Their perspective is that the best way to manage the connection activity of their world is to sit in the back seat and let someone else drive them. They may attempt something new such as joining a social network or listening to a podcast, but unless they see this activity as something of value to their work or social lives, they do not adopt it consistently or practically. It's easier for them to observe what is happening around them, occasionally immersing them in the fray briefly but generally opting for the non-participatory stance.

The majority of this segment is classified in the baby boomer generation, with good representation from pre-boomer, Gen X, and Gen Y depending on their attitude toward the value of connection technology. It is understandable that the boomer generation would be most represented, as this group was brought up on technology that was designed for passive consumption. From the middle ages of radio to the birth of television, these media vehicles required more listening and watching behavior to extract the best value from them rather than through interaction. Pre-boomer members generally graduated from the *basics* as a result of "taking the plunge" into the connected world by adopting a technology they deem of value. An example is of grandparents who need to be able to access photo-sharing sites such as Flickr and Picasa, to be able to view photos—posted by their Gen X children—of the grandchildren during their various life development stages. Once again, the behavior is passive in that they will access these sites only if there is a social need or inherent value to be had.

This group has a very good understanding of the application and device capabilities relative to their social and/or business circumstances. The key difference between this segment and the basic connectors is their optional choice to participate. This group chooses not to consistently adopt or prioritize the use of a new technology due to their passive behavior. They have a strong preference to sit, listen, and wait before acting, opting for someone to contact them via a voice message or some written form.

The device that personifies this group is the base-level mobile device. Their preference is to stay connected via a device that can sit comfortably in their pocket or purse that they can choose to interact with—or not—at their leisure. If a call comes in that they choose not to take—a persistent colleague or the chatty sister-in-law—a favored application is voice mail that manages the connection. Voice mail provides the ultimate screening of mobile phone calls and provides options to either call back or ignore for these passive users. Similarly, applications such as e-mail and instant messaging are preferred. E-mail allows the option of not responding to the message sender or, at the very least, responding at the recipient's own good time. For instant messaging, although *passives* may be listed as contacts, they are rarely active online and respond only when necessary based on their personal agendas.

Passives usually have "visible" connections numbering fifty to one hundred. They may also belong to at least one online social or business group. They are very selective in their choice of friends and the perceived association others would have about their circle of friends. To be selected by *passives* to join their online network is tantamount to membership in an elite group, as long as one is willing to accept a high level of non-responsiveness to inquiries, requests, and interactions from them. As with *basics*, it is ironic that this group may have many invisible contacts, such as in the case of Diana who, given her high profile in both the corporate world and her community, would have many stored contacts. However, these contacts are highly moderated for relevance, usually by an assistant or significant other. They also possess a good social group and may interact with them regularly, but *passives* manage their specific connections at their own rate and pace. Their process of staying connected to people is on an ad hoc basis and usually prompted by other, more connected friends and family.

A famous *passive* connector is American homemaker, Martha Stewart. Stewart boasts a rich resume of achievements through her Living Omnimedia multimedia empire, which includes magazines such as *Martha Stewart Living,* a television series *The Martha Stewart Show*, a radio program *Martha Stewart Living Radio*, and the best-selling book *The Martha Stewart Home Keeping Handbook.* According to a blog post by Nicole Ferraro in the *Internet Evolution,* Stewart announced her intention in October 2007 to create a social networking capability called "Marthapedia," which claimed to "encourage users to generate and add content" to her existing web site. The initiative was doomed to fail when an *Advertising Age* article reported "before posting anything user-generated, editors at Martha Stewart Living Omnimedia will check to see if the public's ideas are better than their own." It seemed that Stewart, despite her multimedia communication pedigree, preferred to take a more passive, one-way interaction with her devotees.

* * *

The *Passive's* attitude toward technology is as a "late adopter." Due to their "lean back" behavior, they wait until a technology device or application is well established in either the market or adopted by their circles of influence before they take the plunge. Although not as conservative as *basics*, they have an understanding of technology and are able to learn and use relevant connection technologies. They have a greater propensity to seek to understand the workings of an application, with a cautious ear on the benefits to them so they can decide whether to adopt or not. Their preference is to spend time on connection devices and applications that are user friendly and of interest. *Passives* are challenged with rate of adoption. They need to be sold on the value of using a technology and take their time to assess and experiment with the technology before starting to use it. Even after adoption, they may decide to use it passively or never access the application again.

A story that illustrates passive connection behavior comes from student Sam. Sam's girlfriend happens to be my daughter, and they came over to our house for our weekly dinner evening. After dinner they both parked themselves in the lounge room and my daughter logged on to her laptop computer. Being the inquisitive type, I asked her what she was doing. She told me that she was organizing Sam's twenty-first birthday party and was logged in to Sam's Facebook profile so she could personally contact and invite *his* friends on *his* behalf. So I asked the obvious question, *"So why isn't Sam inviting his friends to his own party?"* I expected either a he's "too lazy" or "can't be bothered" response. Instead I got a practical and efficiency-based one. *"Look, Dad, I can just access his profile, send a note to all his friends from him (so they don't have to be my friends) inviting them to his party, and we can manage all the RSVPs through his profile so it doesn't clog up* mine." My daughter is an *active* connector, which you'll read about in a later chapter.

Passive connectors are passengers when it comes to connecting. They do not prioritize the use of it in their daily lives unless convinced of the practicality. They will try different applications if prompted and ride the wave until they either get bored or find no value. Similar to *basics*, they know at least one person in the other connector profiles. This is usually someone who is online savvy who they rely on as their "connection agent" to assist them in managing and interacting in the online world on their behalf. *Passives* are playful, sociable, but noncommittal toward connection. They particularly enjoy "surfing the web," skipping from web site to web site browsing what may be of interest to them without actually engaging or buying anything. They also use sites such as Wikipedia to find out more about a topic but do not post content themselves. They may also read blogs of interest that may have been recommended by friends but rarely if ever post comments.

Pooja works for a global information technology company and is someone I mentored for some time. She contacted me one day to inform me that she'd decided to spend two years in India on assignment and although she had family there, she didn't have many business contacts, apart from internal company colleagues. She wanted access to more local contacts to develop a profile with key recruiters in case she wanted to find another role longer term. Although she had registered on online networking site LinkedIn some time ago, she had completed only a limited profile, which didn't reveal much about her skills and experience. With such a "passive" profile, it would have been difficult for potential prospects to not only locate her, but, once found, they would have a limited view of her business skills and experience.

* * *

Being a *passive*, Pooja advised me that she was leaving in two weeks and wanted to begin the process of creating a more visible online persona and getting more connected to key contacts that would benefit her longer term. I coached Pooja on what she needed to include on her LinkedIn profile, which essentially was an accessible version of her existing resume. To get her profile to 100 percent completion, she needed to be very crisp in summarizing her best attributes and to choose the right "tags"—key words that would link searches for those words with her profile so she could get found more easily. She also needed to make formal requests to her referees to post publicly visible recommendations of her work so they could appear on her completed profile.

We developed a plan to target relevant connections by leveraging my larger base of direct contacts and by joining key groups such as Friends of India and the India Leadership Network. I reminded her that this was not a "one-off" activity and that she needed to develop the discipline to access Linkedin regularly to continually invite and accept connection invitations, begin to participate in discussion forums by not just reading information but posting her own thoughts and opinions about the content, and, most importantly, persist with it.

Pooja started the process and had posted a completed public profile within days. However, being a *passive*, she found it more difficult to regularly access the application to continually build her connections. This behavior required some "pushing," which resulted in Pooja increasing her direct connections from twenty-seven to one hundred-one within forty-five days.

Like *basics, passives* are willing to adopt applications if they can be convinced of the value and, more importantly, see the results of their efforts. As they possess an understanding of the technology, they need only to overcome their reluctance to persist with the effort and to see some short-term results before they develop the capability and willingness to adopt the application. Situational and circumstantial factors play a significant role for *passives*. The key contextual consideration is the importance of friends, family, or peer groups. Much of the behavior of this group is determined by how others use technology relative to interaction and collaboration.

Diana is a busy executive but still has a need to "stay in touch" with colleagues, staff, and family. Using a PDA device such as a BlackBerry illustrates her need to retain a connection to her social and business clusters even in an ad hoc method such as monitoring e-mail on a portable device. Similarly, Loretta's interest in connection is stimulated—and activated—by being prompted to act by an invitation or piece of information as is her curiosity in the mass adoption of social networks. Sam will stay passive until he can see the value of being and staying connected online to his support group.

Passives are also capable of adapting to another connection profile based on situation or circumstance. Pooja needed to get connected fast to business people in India to create a visible online and offline persona and became more selective or active when faced with a context that required her to. On the other hand, *passives* are also capable of regressing to *ba-*

sics if they are not stimulated by their friends, family, or peer groups to stay regularly connected, even at a passive level.

* * *

Socially, the challenge for *passives* is to "keep up with the Jones'" in their social and business groups. It would be easy for them to become disconnected from their groups due to deprioritizing connection activity and their "lean back" behavior toward staying connected. Although their preference is for more offline interaction, they may "miss out" on social and business information and events if they choose not to keep pace with the level of connection activity that may be occurring online. From a business perspective, *passive* workers will struggle to stay relevant in an increasingly connected business environment. The need to actively participate in connection activity with customers, prospects, partners, suppliers, and employees that will increasingly occur due to greater efficiency through applications such as online networking, instant messaging, and blogging means that *passives* may find it difficult to remain competitive for jobs in a corporate environment. As businesses find that their customers are becoming increasingly connected, then the employees who sell, market, and service those customers will need to have access to those customers socially, behaviorally, and virtually.

* * *

SUMMARY: Passive Connector

* Attitude—lean back, passenger, late adopter
* Perception—Conceptual
* Contacts—some to multiple, but usually invisible
* Connections—between 50–100 visible and knows a connection agent
* Connections technologies—1–4, but selective for application only
* Direct time spent—minimal
* Examples—executive Diana, manager Loretta, student Sam, professional Pooja, and homemaker Martha Stewart

CHAPTER SEVEN
SELECTIVE CONNECTORS

The internet had given her voice back. And what a wonderful voice! A woman born in 1898 was now an internet pioneer... She even started getting fan mail from users who sought out her advice. Oh, did she love that! Blogging made her relevant again. She had so much to offer, so much guidance, inspiration, knowledge, and experience to share.
— Bill Shafer about Ruth Hamilton, aged 109, growingbolder.com, August 2008

* * *

Olive Riley, Australia's claim to being the world's oldest blogger, had a compatriot centenarian. Ruth Hamilton from Orlando, Florida, in the United States was three months shy of her 110th birthday when she died in January 2008. Her life highlights included marriage to an American Major League baseball player and being one of the first women to host a radio show in the United States. In 1937, during a visit to Europe, she came face to face with Adolf Hitler. Later she entered the political arena and became the first woman elected to the New Hampshire Legislature. During her later years and with the express purpose of relating her life experiences for the benefit of others to read and learn from, she produced and posted video blogs on *GrowingBolder.com*, making her the oldest person ever to blog and be a member of an online social network.

Melissa is a vicenarian at twenty-one years of age and a hairdresser. When her friends began using online social networks to stay connected, she decided to try her hand at it. She posted a Facebook profile and started to use some of the applications. She quickly learned that it wasn't something she wanted to spend a lot of time on so she began to use it sparingly and selectively. Her boyfriend is a *basic,* which is another reason why she didn't cruise the network regularly. She would access her Facebook page when she wanted to invite her close friends to her social network but accepted invitations only from people she knew, and, more importantly, friends she wanted to stay connected with. She would also post "only the good photos" from family events and parties she attended and although her preference was to communicate on the phone or face-to-face with friends, she knew she had to monitor her online party invitations—as her social group rarely sent hard copy party invitations anymore—so she didn't miss out on any group events. Although she had a keen interest in music, she bought an iPod after everyone else told her how good they were. She has more than 220 contacts on her mobile phone and uses text messages to communicate, primarily to friends and family, mainly to keep her costs down.

Richard is a tricenarian in his late thirties and senior sales manager for a global information technology company. I once described his business expertise in the following six words: Understands Business Sales People Management Discipline. Richard has always understood the importance of two things—relationships with customers and friends and managing time in a disciplined way. He managed a team that covered a sales territory of around six hundred small and medium business customers. To allow him the time to spend managing and coaching his team while spending relationship time with key customers, he had to be disciplined in his approach to how he spent his time at work. He decided to choose a core set of customer decision makers, business partners, and peers with whom he would establish close relationships, who could assist, mentor, or partner with him for commercial business benefit.

Richard's appeal to his staff is his ability to blend hard work with having a good time. He is sociable by nature and enjoys mixing with colleagues and friends. Richard had 130 connections on the online business networking site LinkedIn and 145 connections on the online social networking site Facebook. In addition, Richard lives in a coastal town that requires a commute of almost ninety minutes a day to and from the sales office. To maximize his productivity, he set himself up to make and take business phone calls through his in-car-enabled mobile phone. He also downloads the latest business podcasts onto a CD or iPod, which he listens to while driving to keep abreast of business trends and insights. He's arranged with his management to work one to two days from his home, having set up a home office with internet broadband and teleconferencing capabilities.

* * *

Despite a gap of more than seventy years between Ruth and Melissa, the four profiles are *selective* connectors. *Selectives* have a good understanding of communication and connection applications and choose to apply them to their daily lives. They have a RELATIONAL interpretation of connectivity and visualize connection as a handshake, a kiss, or a lightning bolt, revealing a personable approach to using connection technology. They manage connections in a planned way, preferring to selectively retain connections relative to the situational context. They are proficient with technology as they have either learned to evolve with it over time—such as Ruth's experience with the early years of radio and television—or applied it directly to their circumstance, as is the case with Melissa's and Richard's adoption of mobile technology to suit their particular needs. It's never too late to utilize a connection technology for that specific purpose.

Olive, who passed away in July 2008 at the age of 108, shared her thoughts and life experiences of surviving world wars and economic depressions, raising three children as a single mother, and working in diverse occupations including as a cattle station cook and a city barmaid. She posted seventy entries on her blog, boasted an international readership in the thousands, and had regular interactions with people from Russia and the United States.

The *Selective's* perspective is to manage connection activity as effectively and efficiently as possible. This means that the message or task needs to be relevant to the technology being used. Ruth and Olive both found that written and video blogs of their rich experiences were the best form of not only communicating, but also connecting to readers and viewers with their life stories, attracting audiences from all over the world. Richard is exceptionally adept at applying the technology selectively to his needs by maximizing his travel time with a mixture of business relations and self-development time.

This profile has almost equal representation across the generations with a slightly higher representation from the Generation X group. This generation encompasses individuals who would be in professional and/or management business roles at this stage of their working lives. They would essentially be "time poor" due to the requirements of potentially demanding and competitive positions within their businesses, necessitating some form of time-management system to cope with the balance of work and life. This discipline would extend to the management of both their business and social networks as is the case with Richard. Pre-boomers would be the group most interested in maintaining some level of connection with other generations.

A growingbolder.com blog post on aging called *The Lesson of Ruth 1898* cites an Evercare survey that polled one hundred Americans turning one hundred and older in 2007 about their habits and interests. A key finding of the survey was that almost 25 percent of the centenarian respondents purchased a music CD, 6 percent spent time on the internet, and 4 percent said they listened to music on an iPod. The author summarizes the report stating, *"The results confirm that the way to stay alive is to stay involved. It turns out that enthusiasm for new experiences may be the fountain of youth."*

Selectives have a learned proficiency with connection technologies relative to their social and/or business circumstances. The key difference between this segment and *passives* is their ability to match a specific connection vehicle or application to either the message they wish to convey or the task that needs to be performed. This group selectively chooses to adopt or manage the use of a new technology relative to the context of their situations. They have a strong preference to connect to people they specifically select in a proactive and planned manner. The device that personifies this group is the advanced-level mobile phone. Their preference is to stay connected via a device that they choose to interact with relative to the person, situation, or message. This group is most likely to activate a caller identification function on their mobile phone where they can see who is calling them via a recognizable phone number prior to taking the call. Caller identification provides the ultimate management application for mobile phone calls and provides either a binary "take it or leave it" action toward any incoming calls that suits the *selective*. Similarly, applications such as online networks and blogs are also favored. Online networking provides the capability to carefully choose who to connect with from the choice of friends, family, and colleagues and, conversely, who to politely "ignore." Blogs also allow *selectives* to be picky about which blog to subscribe to and participate with on some level, versus others of no interest whatsoever.

Selectives usually have "visible" connections numbering 100–150 connections. They may also be a member of two to three online social or business groups. This segment is über selective in their choice of friends, family, and colleagues. To be accepted into a *selectives* online network is to be a member of an exclusive club, with minimal but rich interactions. Unlike *basics and passives, selectives* have a discriminate group of visible contacts, viewable only by the members of the network. These contacts are carefully chosen and managed directly. They enjoy a thriving social group with whom they regularly interact online, but preferably offline. Managing their online connections allows for greater time to be spent with friends, family, and colleagues in the preferred face-to-face or phone interaction, in line with their message, task, or agenda.

A famous *selective* is one of the world's most renowned geeks, Bill Gates. Gates had set up a Facebook profile—after Gates' Microsoft had invested $240 million for a stake in the social networking group. Several reports then stated that he closed his Facebook account in February 2008 citing that "he was getting tired of sifting through thousands of friend requests." Shortly thereafter, he created a profile on the online business networking site LinkedIn, which features privacy controls that enable users to block connection requests. As of October 2008, Gates' LinkedIn profile boasted four connections. Shortly after joining, Gates posted a question on the site's answers application, which allows any user on the network to ask a question that can be answered by any other user, which at the time was a potential twenty-five million business professionals. The question posed was designed to solicit ideas and discussion on *"How can we do more to encourage young people to pursue careers in science and technology?"* Within two months, the question generated 3,566 responses, which no doubt Gates' staff is sifting through for implementable initiatives.

* * *

The technology adoption profile for *selectives* is the "early majority." Although they are slightly more conservative in outlook on technology adoption, they are open to new ideas, methods, and applications that will permit them to develop more efficient and effective work and life management systems. They have a propensity to take the time to learn how an application can benefit their specific agenda. Their preference is to spend time selectively on connection technologies that get the job done in the most effectual manner.

I happen to be married to the person who I must credit with discovering the term *selective* connector. Here is her story:

Kerry has always been "selective" in her social relationships. We have been married since 1985, which would have to define a person who is both loyal and selective! She has actively retained a set of past and present work friends who have supported her through good and bad times. While she is willing to correspond via phone and e-mail with other colleagues of her past who try to reach her, she makes a regular and sustained effort to maintain connections with the "core group" of friends in her social network. As a part-time university student who completed her bachelor's degree in 2008, she also needed to manage interactions with other students in a time-effective way. She managed this through pre-planned discussion forums using instant messaging with her peer student group on relevant subjects. She also managed her informational needs in a structured fashion using search and storing techniques for research conducted for numerous assignments and essays completed over the years.

While all connection profiles have a contextual aspect to their behavior, *selective* behavior is most likely to be directly determined by the situation. *Selectives* pride themselves on maximizing their connection productivity by using disciplined applications and devices to meet their objectives. However, once they have adopted a technology, they are effectively engaged. This engagement outlook ensures that they prioritize technology as an enabler to achieve premeditated and required results, convincing themselves of their practical application. They have no fear in trying different applications but have a short level of patience in getting to a beneficial result. Due to their social nature, they align themselves to people who can aid or contribute to their connection management.

An example of this behavior is Ned, the owner of an outplacement business. I ran into Ned a few days after providing him and his management team a presentation on the commercial benefits of online business networking site LinkedIn. After the presentation, he logged on to the site to search for a colleague he worked with in England and the United States over twenty-five years ago but had lost touch with. He'd been trying to track down this individual by Googling him but the results showed only his past information, such as press articles and media reports with no current information. His LinkedIn search immediately resulted in a posting with his full profile, including a photo and contact details for his New York office. After a quick verification e-mail, the pair connected electronically shortly afterward. Ned then arranged to visit his previously long-lost pal in the flesh on his next planned business trip to the United States.

Selectives are engaged, committed, and laser-focused. They relish connection with the web but on their terms to avoid time wasting and perceived inefficiency. They are specific about their use of search engines in determining what they are looking for and once they've hunted it down, they usually leave. Their online purchase behavior is to search for an item on Amazon or eBay, hunt down their desired item, complete the purchase process, and then exit as quickly as they entered. They may decide to browse but only for items related to their specific interest. They inevitably subscribe to blogs of interest and will post pertinent comments if applicable to their circumstance. *Selectives'* mobile usage behavior is advanced. The device they carry will most likely be the latest mobile, on the fastest mobile network, with the most advanced communication features. The functions most utilized are voice calling, text messaging, and Bluetooth to predominantly connect selectively to other users and devices. They use their mobile phone camera to take photos and send these to trusted individuals in their social group or post only the most appropriate pictures on their chosen online networking sites for their collective connections to view. Generally they have a self-determined mobile ring tone that reflects their personality. In many cases, *Selectives* have their phones turned on and at the ready although in some cases, and as mentioned previously, they rely on caller identification to screen incoming calls.

* * *

Selectives have the capability to act decisively to change their connection behaviors relative to their contextual circumstances and to achieve specific outcomes. I met up with a former colleague, Peter, who had been retrenched from a large telecommunications company. Peter had just arrived back from his honeymoon and thought he was settled in his role with the company. The retrenchment came as a complete shock and he was looking for advice on how to engage with recruitment agencies—and fast. Upon posting Peter's profile on the online business networking site LinkedIn, he had managed to connect to around one hundred people in his business "inner circle." I advised him to conduct a search of the top recruiters in the land who had profiles on LinkedIn and specifically target them to connect with him. Within a few days, Peter had 240 connections, most of them being recruiters in the city he worked and lived in. Within a few weeks he had arranged interviews with the key players and was well on the way to securing a new position to support his family.

Like *basics and passives*, situational and circumstantial factors are paramount for *selectives*. The situational consideration could manifest itself by a person, an event, or group. Much of their behavior is determined by how they choose to use connection technologies relative to how they interact with people and information. Both centenarians Ruth and Olive had rich stories, life experiences, and lessons to relate to an interested world. They chose an immediate technology in the form of written and video blogs to bring their messages to life for all to view. Peter needed to establish a recruitment connection to enable him to find a job as quickly as possible. This meant a change in his previously closed approach to business contacts to the more expanded and practical targeting of recruitment stakeholders that could assist him in achieving his employment goals. Conversely, *selectives* are also capable of regressing to a *basic* or *passive* connection profile if they don't find the time to manage their connection activity.

* * *

Socially, the challenge for this group is achieving a balance between a closed group of contacts and exploring new connections. *Selectives* could remain in their cloistered group without experiencing the opportunity to expand their social and business horizons for their own benefit. They may have to learn the lessons of Gates' crowd-sourcing question and Peter's targeted recruiter connection to be more effective in achieving shorter-term goals. Like *passives*, they could invariably "miss out" on social and business opportunities if they choose to remain in a closed network of contacts.

From a business perspective, the *selective* worker can enjoy rich and deep associations with customers and partners they choose to maintain relationships with. The challenge will occur when the situation changes—a move to a job with a new set of management and peers or a sales territory with an unfamiliar set of customer accounts and stakeholders. Unless *selectives* react positively to a new environment by applying the same connectivity management disciplines, they may struggle to be as effective and efficient. As is the case with Richard's adaptability to a mobile office environment for greater productivity, *selectives* need to be able to move swiftly when faced with the challenge of expanding an exclusive network.

* * *

SUMMARY: Selective Connector

* Attitude—proficient, engaged, early majority
* Perception—Relational
* Contacts—some to multiple
* Connections—100–150 visible, knows who to connect with for social or business benefit
* Connection technology—multiple, selective devices and applications
* Direct time spent—selective
* Examples—centenarians Ruth and Olive, hairdresser Melissa, sales manager Richard, director Ned, professional Peter, and technologist Bill Gates

CHAPTER EIGHT
ACTIVE CONNECTORS

We don't accomplish anything in this world alone...and whatever happens is the result of the whole tapestry of one's life and all the weavings of individual threads from one to another that creates something.
– Sandra Day O'Connor, U.S. Supreme Court judge

Hamish Blake and Andy Lee are professional comedians based in Melbourne, Australia. As aspiring and ambitious comics, they decided to take a break from university study to get into the entertainment industry. They tried their hand at stand-up comedy, presented programs on Melbourne's Student Youth Network radio station, and did some community TV work. In 2005, they were offered an unpaid late-night shift on radio station Fox FM. The station liked what they saw and they went on to host the *Almost Midday* show on Saturday mornings, broadcasting to the key markets of Sydney and Melbourne. In 2006, the "Hamish and Andy" radio show graduated to Fox FM's drive time slot (4:00–6:00 p.m. weekdays) and catapulted to No. 1 after just five weeks. In October 2008, their show was still No. 1 nationally and their combined national weekday drive time and Saturday morning shows attracted two million listeners, making it Australia's most listened to radio show. The "Best of Hamish and Andy" podcast became the No. 1 Australian radio podcast with a monthly average of 240,000 downloads and, at one stage, topped the iTunes charts in their category. In addition, they also had 160,000 fans on Facebook.

Why did they become so popular? Their web site profile declares that *"their chemistry, cleverness, and unique comedic style have earned them broad appeal with a national market."* The clue to their mass appeal could be that Hamish and Andy continually refer to their radio show as the "People's Show." Their show format is a mix of offline—and online—interactivity with their crew, guests, regulars, and, most importantly, their audience on current, everyday, and usually funny topics. A typical show usually starts with a thought or an insight that both have a perspective on, which they share with their listeners. They immediately invite their audience to comment by either calling in to the show talkback-style or sending an SMS or an e-mail to air their views on the topic—which are broadcast live or later, once the feedback has been collated.

Their regular segments include "Blast from the Past," where they call a random phone number and try to convince an unsuspecting person that they went to school with them and the compelling radio visual "Pants off Friday," which reflects the sentiments of their commuting listeners as they make their way home at the end of a busy week.

In 2008, they conducted broadcasts from war-torn Afghanistan, the Beijing Olympics, and while trekking across Australia on their "Caravan of Courage" pilgrimage. They convinced a potato chip manufacturer to launch a gravy-flavored variety, which won a poll of their audience's most desired potato chip flavor and launched it as the "People's Chip." They participated in a party marathon by attending forty-two parties in Sydney, Melbourne, and Brisbane over a fifty-four-hour period, selectively accepting party invitations from their audience. The show web site is a template for a successful Web 2.0 implementation. Visitors can listen to selected highlights from shows they've missed via podcast or minicasts, watch some of the duo's antics on video and in photos, and input into the show by commenting on blog posts or features or by contacting the show directly via e-mail. People can also access their Facebook page and become one of their many fans.

My daughter Rachel is a third-year university student in her early twenties. She balances her full-time studies, a part-time job, and an active social life that includes partying and social basketball. Her familiarity and regular use of technology from a young age positions her well to manage the demands of her university course. The course information for her areas of study is all accessible—anytime, anywhere. The university site includes a course notice board that regularly details important course information for students, downloadable lecture material in audio and presentation formats, and electronic discussion forums to generate dialogue between students and staff about course topics. Her course research, assignments, and notes are developed and stored on her internet-enabled laptop, with her course essays and reports usually submitted electronically. Her positive attitude to the use of technology also assists her to manage her social circles of family, school friends, work mates, and university contacts in a disciplined way.

She uses Facebook to maintain contact with her two hundred plus connections collectively and admits to using it to keep her "finger on the social pulse." On Facebook she can e-mail her contacts, leave messages on their walls if they aren't online, check out photos of various events posted by others, and post some herself. Most importantly, she keeps herself updated with the latest happenings within her network via her home page news feeds and status updates. She relates the story of one of her friends, Chris, who is in an on-again, off-again relationship with Kim. So as not to embarrass herself by asking if they're still together at a point in time, she accesses Chris's—and Kim's—Facebook page prior to any conversation to check their relationship status, in their own words. As it's updated regularly, it is an accurate indicator—at any given moment in time—as to the precise status of their relationship.

At sixty-five, Maurice is semi-retired and first started using computers in his pharmacy business. The attraction to the technology was the efficient way that medicines could be catalogued, accessed, and searched via a simple database. In the early 1990s, while running his small business, he bought two desktop computers and three dot matrix printers for just under $20,000, as he soon discovered that there were other effective applications beyond the database—such as spreadsheets and word processors—that could make him and his business more efficient. Since the mid-nineties, he has been a regular user of e-mail and mobile phone technologies, which he uses to communicate with friends and relatives who reside both locally and overseas. His beloved daughters have now married and moved out of the family home. He has no hesitation in joining his children's online network of choice when they decide to start a family. He knows that this will be the most efficient medium to stay connected to his kids through content updates, photos, and videos, no matter where they live. Ultimately, he knows that it will be the primary interface with his grandkids.

* * *

Maurice, Rachel, and Hamish & Andy are *active* connectors. *Actives* possess a profound understanding of the effectiveness and efficiency of connection technology—and use it proactively to their benefit. They have a PHYSICAL view of the word "connection," and, when prompted, visualize connection as bridges, chains, wires, switches, power points, phones, and even sex. They manage connections by being assertive in maintaining contacts, prioritizing relationships, participating in conversations and events, and regularly seeking input and feedback. As their name suggests, they actively seek interaction with people, information, experiences, and ideas as they perceive this to be of value, both socially and commercially. Their outlook is to manage the agenda by leading the connection activity of their world. They embrace most aspects of communication tools and social media. They are members of social networks, regularly participate in interest groups, and, invariably, are bloggers themselves. They use the technology regularly, practically, and consistently. This group has a tendency to be an early adopter of technology they envision could be of material value to how they personally and professionally operate. Gen X and Gen Y are best represented in this segment followed by boomers and pre-boomers. Some of the pre-boomers and boomers represented in this group graduated from the ranks of *passives* and *selectives*.

Some boomers like Maurice were early adopters by attitude. He took the risk to invest in personal computers early on as well as dot matrix printers, the best printing technology during its day. For Maurice, typing prescription information into a computer database that automatically knew to print a personalized prescription label that would go on the medicinal packaging beat having to do it manually on paper or even using a typewriter. He realized that the pieces of information he needed to match such as medical information and the medicines he was dispensing were linked. He was astute enough to figure out that the most efficient way to connect them was to handle the information once via a database.

Actives comprehend the value of connection technologies relative to their social and/or business circumstances. The key difference between this profile and *selectives* is that they behaviorally drive connectivity to suit their needs and, more importantly, see a range of connections to people, information, experiences, and ideas as opportunities. This group chooses to consistently adopt or prioritize the use of a new technology due to their active behavior. They have a strong preference to initiate interaction by taking the lead in stating, commenting, acting, and seeking input from their connections via any medium that suits their participatory style.

The device that personifies this group is an internet-capable mobile device, such as the iPhone. Their preference is to be in an "on call" state via a device that can connect them to people and information instantly. Their mobile device will usually be within arm's length 24/7, with calls answerable regardless of who the caller is. When they're busy or otherwise indisposed, they have a preference for text messaging to voice mail due to the "handle-it-now" capability of the medium. Productivity and internet-enabled applications such as Google for search and Facebook for social networking are preferred by *actives*. Google permits a number of information (web, news, reader, images, etc.) and location (maps) search options. Facebook is the social platform that keeps them connected to friends and acquaintances as much as to the "happening" events of their social network. With the ability of internet-enabled mobile devices to incorporate the basic functions of mobile voice and messaging combined with the web capabilities to access search, e-mail, and online networks, these devices are made for *actives*.

Individuals belonging to this segment have between 150 and 300 connections, are likely to be connected to one to two online networks, and are members of two to five online social or business groups. They are more likely to have many contacts and proactively endeavor to connect to as many of them using online networking tools as possible. Once you've accepted an invitation to connect with them, they will keep in regular communication by soliciting information and ideas from their connections and asking you to join interest groups to maintain a connected relationship. *Actives* are happy to share their connections with others— and for their networks to connect to each other and share information—as they view this as an opportunity for collective benefit. Hamish & Andy proudly exhibit the names and faces of their 160,000 fans on Facebook and encourage them to generate conversations and ideas that the show producers regularly review to see what their audience is interested in and what they want to hear more about. By "listening" to their audience in this way, they position themselves well to retain their audience's interests and keep them as loyal fans.

The most recognizable *active* is U.S. media mogul Oprah Winfrey. Her appeal and reach through her books, television shows, and web sites attract many people to connect to her message and initiatives. Her impact on both local and global communities was best summarized by Michelle Obama in *Time* magazine's 100 Most Influential People in the World in May 2008. *"Oprah has developed and nurtured a relationship with her viewers and readers built on the recognition that there is more that unites us than divides us—that our shared experiences in work, life and love, in family and community, in our hopes and dreams, know no barriers; that regardless of race, gender, socioeconomic status or hometown, we are our brothers' keepers, our sisters' keepers."*

The prevalent attitude to connection technology for *actives* is "early adopter"—proactive and influential in their social communities and open to new ideas. Although they may be skeptical at first due to their assertive behavior, they take the time to learn and understand the practical uses of connection technology and instinctively apply them readily to a situation or circumstance. They possess an astute propensity to identify how an application will benefit them or the connection process quickly. This attribute makes them perfect connection agents for *basics* and *passives*, as is the case of Geoff the taxi driver and his son, who assisted with the payment of his boat registration payment, and Rachel, Sam's girlfriend, who effectually manages his connection activity for him.

Their preference is to plan and prioritize time spent on relevant connection devices and applications. They regularly review and will experiment with other technologies of interest as a form of process efficiency improvement. Once they adopt a connection technology, they will work it to their benefit by immersing themselves through regular participation.

As much as *passives* are passengers, *actives* are drivers. They are assertive about their use of technology, prioritize its regular use in their daily lives, and actively participate in the various communities they belong to. For some, it may take a little convincing to adopt multiple applications, but once they understand the value they not only catch on quickly, they drive it to their personal and commercial benefit.

A story that personifies *actives* comes from Michelle Zamora, a thirty-something senior marketing manager at a global IT company and a working mum. This excerpt from her blog post is reproduced with her permission.

I have to admit that I was one of "those people." You know, the ones that think Facebook, Twitter, blogging, and oh golly, YouTube were for mindless entertainment—for people who have too much time on their hands. I simply just did not get it. My initial perceptions of social networking really stemmed from the use of the word "social." The Australian culture tends to use the term social to refer to leisure pursuits, such as gatherings with friends over a weekend BBQ, as opposed to the development of communities of like-minded people. So to me social networking was something you would do with your time off work—entertainment at best. It seemed the art of holding a conversation was replaced with the unique ability of biting one another with their Vampire Alto-Ego on the virtually attached Facebook. OK..........not for me.

After the birth of my third child, I commenced working from home more often. A work benefit I appreciated and enjoyed, a benefit that enabled me to work productively whilst being accessible to my children. But slowly I became all too familiar with the isolation and lack of community that was being created in my world. By chance, the opportunity to work on a project arose, which would require me to better understand social networking. And so began my journey—a journey I thought that I would, well, blog about. I started by looking at my professional networks. So I can use LinkedIn to build my network. OK got it. Now what. With caution I stepped into the world of blogging. Initially I simply added comments to existing blogs—I stopped just being a silent observer and began participating in online conversations. Soon this was not enough. I needed a creative outlet and found that I had something to say. With trepidation, I commenced my own blog. And over time realized I could promote and market my blog through links in my social network. So now time to look at Facebook—reluctantly I created an account—and well, have become a Facebook Fan. I can build my personal network, I can join groups of interest and meet new people with common interests in marketing, branding, innovation, parenting—and these contacts just maybe will become part of my professional network in the future. Wow, who would have thought?

I met @Wonderwebby via instant messaging. She encouraged me to take a look at Twitter. At first glance it looks like a, well, time waster—entering in what you were doing. Now really, who has time to write about what they are doing—I do not have enough time living life. Then I discovered a world of thought provocation—access to people who would challenge me to rethink my opinions—who would allow me to understand how others are using their internet and social networking strategy, the latest in branding and innovation across the globe, and also have a little bit of fun. Wow, so I am not the only parent still awake at 1 am with kids who are determined to keep waking up on the hour. Thank goodness I am not alone. Sanity retrieved! So now I can tweet a thought, generate ideas and a greater understanding of common dilemmas and information requirements, and partici-pate and lead blog conversations. So what do I now think of social networking? Social networking is simply a way to collaborate with people on a global scale, with the tools you choose to use. I can do that.

* * *

Actives like Michelle have developed the capability to use connection technology to their benefit. They drive the use of connectivity to suit their circumstances by initiating inter-actions. They proactively prioritize their connection activity. The opportunity for businesses that employ people like Mi-chelle is to leverage this connectivity drive for commercial gain. *Actives* make great sales professionals and managers, as they have the capability to proactively introduce themselves, their companies, and their ideas to prospects and customers, develop strong relationships with existing retention clients, and continually build and manage their connections base to maintain business relationships resulting in more sales trans-actions for the organization.

This also applies to marketers who are able to do the same in establishing relationships and connections with key stake-holders in their target markets. The business implication for *actives* is that they might take up too much time busily con-necting if a system is not in place to control or manage their connection activity. While they have a willingness to try different technologies, the downside is that they could become unproductive if they cannot identify the personal and commercial value of connecting expediently.

* * *

SUMMARY: Active Connector

* Attitude—understand technology, early adopter, and driver
* Perception—physical
* Contacts—multiple and visible
* Connections—150–300 visible; member of 2–5 groups and 1–2 networks
* Connection technology—multiple and relative to application context
* Direct time spent—significant
* Examples—comedians Hamish & Andy, student Rachel, retiree Maurice, Michelle the working mom, and media mogul Oprah Winfrey

CHAPTER NINE
SUPER CONNECTORS

*The ability to relate and to connect, sometimes in odd and yet
striking fashion, lies at the very heart of any creative use of mind,
no matter in what field or discipline.*
– George J. Seidel

* * *

Barack Obama could be the most astute connector on the
planet. He knew from the start of his campaign that to become the president of the United States in 2008 he needed
to be where his constituents were. In his mind, he didn't
need a study by the Pew Internet & American Life Project,
conducted in April/May 2008, to confirm what he already
knew instinctively. The study found that one in six American
adults were using the internet to keep up with the election campaign, via viewing political web sites and reading
e-mails and text messages. The study also showed that the
key stakeholders of Senator Obama's constituency—young,

black, and affluent voters—showed high rates of online news consumption.

According to a Reuters report on the study, *"Democrats were also more likely to use social networks, watch online videos, donate to campaigns online, sign up for campaign-related e-mail and bypass news outlets to get information directly from campaigns."*

To win this election, he not only had to communicate to these voters, he had to engage them with his message and proposed policies. More importantly, he needed them to continue the conversation amongst themselves and influence others in ways they were most comfortable with. He called this "interconnectedness." By October 2008, Barack Obama "owned" the online networking space. He boasted 740,000 MySpace friends and over two million Facebook fans. If you didn't sign up as a fan, you were regularly served up advertising on your Facebook home page encouraging you to download his free audio book.

He was ranked the fifteenth-most connected individual on the business online network LinkedIn, which numbered thirty million predominantly North American senior business professionals and executives at the time. He has also raised tens of millions of dollars in campaign funds over the web, mostly in micro-donations from masses of individual supporters. Obama's magic resonated globally as well. In an October 2008 Gallup poll conducted in seventy countries that represents nearly half the world's population, 30 percent of respondents preferred Senator Obama as president of the United States against 8 percent for Senator McCain, with 66 percent of Japanese and Australian respondents preferring Obama to McCain.

Ron Bates is in the people business and a recruitment expert. He relies on having the best information available to him about the executive job market by knowing about upcoming resignations, planned and unplanned organization restructures and redundancies, and keeping tabs on talented candidates who are pondering their next career move. It's his job to be positioned to connect his clients' job vacancies to the right candidate at the right time and provide guidance and expertise on recruitment. Ron is a search consultant with Executive Advantage Group's Silicon Valley office, based in San Jose, California. His online biography states that *"He is actively involved with Board of Directors and Executive level searches for early-stage multinational companies, as well as expanding the firm's Technology Sector specialty group."*

He is also a recognized expert in online networking, personal branding, and building a personal internet presence. He has been a speaker at the Marketing Executive Networking Group, Human Capital Institute, British America Business Council, and Business of Success Radio. By October 2008, Ron had accumulated 43,000 direct contacts on online professional networking platforms. Ron has been referred to as "the most connected man on Earth." On his LinkedIn profile, he boasts that *"There isn't a recruiter in the world who can identify, qualify and recruit Exec candidates better than I."*

Mark is in his twenties and just moved interstate to accept a job as a concierge at a major hotel in a big city. The decision to move was facilitated by his close connection to the multitude of family and friends on the online social network Facebook. He also keeps in touch relatively cheaply by subscribing to the same mobile phone network as his social circle, as it offers free calls for designated groups of users. During his regular home visits, he always seems to have the latest mobile technology, whether it's a new Nokia mobile, the new release MacBook Pro with matching iPod or a state-of-the-art GPS navigation system in his car. He has a resourceful attitude for searching and sourcing information and services, which makes the concierge role a perfect fit.

* * *

Within the extent of their circles of influence, Mark, Ron, and Barack are what I call *super* connectors. Whether it's for social, business, or political rationales, they have an instinctive understanding that utilizing the most appropriate technological tools and applications will help them achieve their goals and desired results. They have a STRUCTURAL perception of connectivity and describe connection as family trees, process maps, organization charts, and networks. They manage their connections in a very disciplined and innovative manner, resorting to the use of connection technology to maintain and grow their connections to people and information. As expert connectors, they strive to actively manage their connection activity in a planned, executable, and measurable fashion so it functions like clockwork. They see any activity that promotes their connection to relevant people and information as productive and worth the time investment. They spend as much time on retaining and maintaining connections as they do in acquiring new ones. They have an "open" view to networking, will connect to anyone who has a mutual personal or commercial interest, and usually make their connections visible to others to encourage interconnection. As experts, they practice, participate, and collaborate at the optimal rate.

The majority of this segment is classified in Gen X and Gen Y, with the boomer generation following closely behind. The rapid emergence of e-mail, mobile technology, and the internet in the working lives of Gen X and Gen Y sets them up to extract the best value from connection technologies more efficiently and effectively than more traditional forms of interaction. No members from the pre-boomer generation are evident in this group. *Supers* have an expert understanding of connection application and device capabilities, and they have an instinctive sense about what works and doesn't work relative to their social and/or business circumstance.

The key difference between this segment and *actives* is simply degree of expertise. While *actives* drive to be and stay connected regularly, *supers* are disciplined, so much so that it is as much a part of their daily routine as brushing their teeth. They don't think of networking as a chore—it's imperative to how they run their social and business practices. This group chooses to manage their lives in this way. They rely on a mix of previous experience, technical knowledge, and gut feeling with regard to their connection activity.

An advanced mobile device such as a BlackBerry or high-end iPhone personifies this group. They must stay connected via a mobile device that is handy to their person. They usually choose to interact with it, at the expense of interrupting conversations and meetings. The term "Crackberry," used to describe people with an addiction to these personal devices, probably referred to a *super* connector. Voice mail is relevant only for missed calls when they are on the phone or when they can catch some sleep, which is rare. Most accessed applications are e-mail, text/instant messaging, and online networks on mobile phones. *Supers* have three hundred or more connections, are active participants in two or more online networking platforms, and are members of more than five interest groups. This group understands the importance of retaining, maintaining, and acquiring connections for personal and commercial purposes. Their connections are a diverse mix of family, friends, colleagues, customers, prospects, suppliers, acquaintances, and people they have been introduced to or have met online. They are potentially global, across industry sectors, and from a range of student, professional, and managerial professions. They are an elite group willing to connect to everyone, as they believe that every connection is of potential personal or commercial benefit.

This group had a solid base of invisible contacts to start with. Barack the president, the most famous of *super* connectors, had a strong support base entering the presidential race, and Ron the recruiter had a database full of contacts from previous search engagements. The difference is that both of them knew how to use connection technology and apply it to retaining and attracting more connections through both direct contact and by association. When I found out that Ron the recruiter was the most connected person on LinkedIn, I was compelled to join his network, just to have the capability on hand to tap into the richness of talent in that group. I also happen to be connected to Barack the president on LinkedIn.

* * *

Supers have an innovative attitude toward technology. They prefer to manage their connections rather than be managed. People like Mark the concierge—with his rapid adoption of new devices —are eager to be the first to try out a new technology device or application even if it's not established in either the market or adopted by their circles of influence. They are the first-to-market people and live on the "bleeding edge." They have confidence in their instinct and experience that chosen devices will work for them, not against them, in achieving their connectivity objectives. They spend the maximum amount of time conducting their research prior to selecting a technology. However, their decision and action process is compressed to quickly progress to test and trial phases. They have a tendency to experiment with different applications and devices before settling on one relevant to their preferences—until something comes along that's better and faster.

Just before my brother's wedding, my wife and I discussed the process of how we would manage the numerous photos we would be taking at the church and at the reception. At previous functions, we would bring along the digital camera, take as many photos as we liked, come back home, log on to our internet-enabled computer and upload selected photos to Facebook so we could keep a perpetual record and share our photos with the family. All of this could be done when we found the time in our busy lives to do so. At around the same time, I had purchased a BlackBerry Bold, which was capable of uploading photos directly into my Facebook profile, with a time and date stamp. I have an eternal photographic record of the last moments of my brother's life as a bachelor on my online social network page. All done in a single movement using a web-enabled device. With my use of technology and number of connections, I'm classified as a *super* too.

Super connectors are efficient managers and practitioners. They exhibit an unparalleled discipline in maximizing the use of connection technology for their benefit. They are innovators who strive to find a better and faster way to achieve their desired outcomes. A positive example of this is Karl, an

American college student in his twenties who participated in a student exchange program at Sydney's Macquarie University. Karl's goal was to absorb as much Australian culture as possible during his yearlong stay, to make as many friends as he could during his time, and to maintain contact. At the end of his program, he accumulated 570 friends on Facebook and is known to "live" on the online social network. He is a member of multiple groups across the globe and is an active user of various social applications. I know this because my *active* daughter Rachel went to university with Karl and is one of his many Facebook friends.

* * *

Supers are ruthless in pursuing connections to their network, sometimes bordering on the intrusion or sending indiscriminate e-mails called spam activity to generate more connections. An example is an invitation I received from a *super* connector to join the LinkedIn network.

*You are currently **not** my 1st Level Linkedin Connection, but I thought I would just send one quick e-mail to you to briefly inform you on who I am and why I'm contacting you. I work in the area of Finance/Management/Consulting as a Senior Financial Analyst. I seek to establish networking relationships with Entrepreneurs, Business Owners, Senior Managers, Executives and Leaders of well-known or otherwise respected companies, either domestically or internationally, in order to expand the development of my career and to further increase my knowledge in the area of finance, accounting, management, business development, real estate, e-commerce, and information technology. I have an open and entrepreneurial mind-set, and welcome exciting & advantageous opportunities.*

First, being a professional networker and connector, I am a Linke-din Guru/Expert with 23,000+ direct connections. I am here to help you succeed too. I wanted to know if you have any questions or suggestions on how to enhance Linkedin amongst all the members, be it in network or in groups. Please let me know. Second, since at this moment you're not directly connected to me, I would be honored if you would invite me to your Linkedin network. I am well-connected on Linkedin and know a lot of people in various industries (we have mutual connections already). As of right now, I don't need anything per se, but who knows what the future holds or what other mutual benefits may arise from it… Let's add value to our development! I am a trustworthy individual, and I would value your privacy greatly. Please see my comprehensive Linkedin profile, detailed CV and 480 recommendations in case you're skeptical. Just click this link for an automatic way to add me. Finally, if you're relatively new to Linkedin or the "web 2.0 networking" scene, here are some good reads about Linkedin and general networking tips (posted weekly) or learn even more about me and see why I am the "King of Networking."

The key contextual consideration of *supers* is the importance of connections to their world. They have a powerful sense of how to connect and an instinctive sense and intuition in creating networks of benefit. Due to this degree of expertise, they can be valued mentors and role models for people who want to commence or enhance their networking skills. They can also behave in ways that make connecting with them seem like an intrusion and can also be a distraction to day-to-day activities. In most cases, *supers* manage their connection activities in a disciplined manner with a view toward maintaining professional connection conduct at all times.

Stan Relihan is a leading Australian executive recruiter, a Top 50 LinkedIn user globally, and is the No. 1 LinkedIn user in Australia. He also hosts *The Connections Show*, a regular audio podcast that puts listeners ahead of the online business curve by reviewing the latest developments in social and business online networking. In each episode, guests from his extensive global network share their experiences and insights on how Web 2.0 tools and resources can be implemented, utilized, and enhanced to create business advantage. Stan sends reminder e-mails to his connections each time he posts a new episode on the show's web site. He receives a share of e-mails back from people who don't wish to receive such notifications, and he diligently removes them from his mailing list. For regular subscribers of the show who are too busy to download each episode, he offers a technology known as "pushcasting" that delivers each episode automatically to people with BlackBerry devices who have signed up for the service. The show currently ranks No. 3 on the list of "Most Popular Business Podcasts"—ahead of both *Business Week* and *WIRED News*, at popular web content aggregator Digg.com, and generates more comments than all other Top 20 shows combined.

Socially, the challenge for this group is to stay "ahead of the pack" at all costs in the connection stakes. They can have a reliance on connecting online only—a passion that can be to the detriment of other social interactions including face-to-face interaction. They are also likely to abuse the privilege of hyper connectivity such as in the case of Corey Worthington. Worthington hosted a party at his suburban home in Melbourne, Australia, in January 2008 while his parents were on a vacation. He sent out the invitation to his party via text messaging, MySpace, and Facebook reading: *"Oh yea party at Mine Saturday 12th Jan. BYO chicks and grog. No knives, rents (parents) will crack it at me."*

The party drew five hundred attendees, which caused subsequent vandalism around the neighborhood. Thirty police officers were called in, including a police helicopter and the dog squad. Police were attacked with bottles and stones, and blamed alcohol consumption and the lack of adult supervision for the party getting out of hand. Following the party, the police commissioner from the state of Victoria wrote an open letter asking young people to be aware of the power of text messaging and the internet.

From a business perspective, *supers* are assertive and ideal candidates for sales and marketing roles. They are extroverts who, if guided in the right way, can make very successful account managers and service relationship personnel. However, if placed in roles that do not take advantage of their connective capability, they could be easily distracted, leading to a reduction in productivity and commercial outputs.

* * *

SUMMARY: Super Connector

* Attitude—connection technology expert, innovator, disciplined manager
* Perception—structural
* Contacts—numerous and diverse
* Connections—300+, member of two-plus networks and member of five or more groups
* Connection technology—multiple
* Direct time spent—maximum
* Examples—Barack Obama the president, Ron Bates and Stan Relihan the recruiters, Mark the concierge, Karl the college student, and party boy Corey Worthington

CHAPTER TEN
CONNECTED GROUPS

Give people the power to share and make the world more open and connected.
— Mark Zuckerberg, CEO and founder, Facebook

Communities are by definition a connection of individuals, gathered due to a common interest or cause. The attitude and behavior of individuals within a group determines its functional effectiveness and what the group is capable of achieving as a single entity. One of my favorite group dynamics is the audience or Mexican wave that's common at sports events packed with people in large stadiums. For those who haven't experienced this global modern-day phenomenon, it usually occurs when there is a delay or lull in event proceedings and the crowd starts to get bored. *Wikipedia*—an information source from the collective global wisdom of individuals—describes the dynamic as *"when successive groups of spectators briefly stand and raise their arms. Each spectator*

is required to rise at the same time as those straight in front and behind, and slightly after the person immediately to either the right (for a clockwise wave) or the left (for a counterclockwise wave). Immediately upon stretching to full height, the spectator returns to the usual seated position. The result is a 'wave' of standing spectators that travels semi-rapidly through the crowd, even though individual spectators never move away from their seats."

Based on personal experience at sporting and selected concert events, the fascinating aspect of this group activity is that only a handful of individuals are needed to trigger it as the group is captive within the confines of the stadium. They also get organized quickly. Most people participate in the activity almost instantly and, as there's no preliminary training to direct people to what their specific role is, more experienced participants—sometimes total strangers—are usually willing to demonstrate to newer audience members how to participate. There is a type of "self-regulation" in the form of sections of the crowd booing and jeering other members who choose not to participate, in most cases cajoling them into completing the action the next time the wave comes around. The result is a coordinated mass of people working in unison to mimic the effect of a wave around the stadium, which includes crowd cheers and applause during and after the activity. The group behavior of a crowd during the audience wave activity is a good analogy to describe how a connected group functions and, when done well, its positive effect on the external environment.

A key connection technology that promotes connected groups is the online network. An online social network provides a platform for building online communities of people who share interests and activities and who may also want to explore others' interests and activities.

A star performer of this technology is the social network Facebook, which began life within an existing group at Harvard University when then student Mark Zuckerberg launched it from his dormitory room in February 2004. The name refers to the paper "facebooks" depicting members of a college community as a visual introduction for new students, faculty, and staff to facilitate getting to know others on the campus. It became an instant hit at Harvard with more than 66 percent of the school's students registering on the site within two weeks. Zuckerberg opened Facebook up to other schools including Stanford, Columbia, Yale, and other Ivy League colleges and schools. Facebook then rapidly expanded to the rest of the world.

Upon joining and posting an individual profile, users can join groups and networks by school, company, region, city, or country to connect with other people, as well as add friends. They notify their friends about what they are doing, thinking, or feeling using a mix of interaction tools such as posting messages and photos on virtual status and profile walls, e-mail, and instant messaging. Users can also form and join interest groups as diverse as Barack Obama's group that had 2,328,227 members three days before the U.S. presidential election in November 2008, U.S. swimmer and multiple gold medal winner Michael Phelps, who has 1.6 million "fans," and the less popular but intriguing "*I Secretly Want to Kill Anyone Who Says the Word 'LOL' (Laugh Out Loud) in Conversation*" group that had thirty-eight members in November 2008.

Members or "fans" of these groups can view and interact individually with other group members, participate in discussion forums, and post pictorial, audio, and video resources. By November 2008, Facebook reported more than 110 million active users and over 55,000 global education, business, and location networks. Within four years of being launched in a small dormitory room, it has become the fourth-most-trafficked web site in the world.

The connected group can be an extension of an existing group such as family, friends, community, business, and global organizations. They can also be formed from new or existing groups with similar ideals, interests, and hobbies such as education, religion, and mothers' groups. They range from being a collection of individuals such as a fans or supporters to large collectives of sub-organizations such as the Olympic family. Their inherent power and influence is determined by their solidarity, which, in contrast to traditional groups, is facilitated and enhanced by the use of connection technology for collective benefit. As a blended mix of connection profiles, they have a group version of the *three Rs of connection*. They also possess a distinct set of characteristics that defines them as a connected group, which I summarize via the appropriate acronym *GROUP*. They have their own unique set of implications, which need to be understood for organizations to better manage the challenges and opportunities they bring to society and business.

As a composite of connected individuals, connected group members inherently exhibit collective *HITS* behavior. They possess *hunting* instincts to track down potential group members and information relating to the interests of the group. They have an intuitive need to *interact* within and outside the group to promote its interests and will *test and trial* various activities, techniques, and ideas that will potentially deliver benefit to the group. Finally, a fundamental reason for the group's existence is usually to *share* information, experiences, and ideas for collective benefit.

Although there is a range of reasons for the existence of connected groups, the importance of sharing to the connected individual has been highlighted in previous chapters as a key reason for their behavior. For connected groups, their very existence as a group is for this express purpose. As a member of a group of like-minded people, the ability to "share at will" becomes easier and more immediate, especially with the use of connection technology such as an online network. An example is attendees to a wedding who form an interest group of the bride and groom's family and friends on an online network. Most of the members would not have met or had much in common apart from knowing either the bride or groom and attending the wedding ceremony. However, a virtual group could form as a result of the event due to a related common interest in sharing individual experiences of the wedding.

On sites like Facebook, for example, it is common practice for a family member, friend, or relative to post digital photos of a family wedding that is visible to the attendee group with the express rationale of sharing the event moments pictorially. This virtual photo album for selective group access serves as a special repository for the numerous pictures taken through the lenses of different attendee perspectives. The specific purpose is to share with others, including known or unknown attendees and even extended to family and friends who may not have been able to attend due to remoteness of location or illness. Photo sharing is a key application on online networks. It's no surprise that in October 2008, there were ten billion photos posted on Facebook at a rate of thirty million uploads daily.

An emerging rationale for groups to become more connected is to enhance collaboration. Collaboration is defined as the process where a group works together toward an intersection of common goals by sharing knowledge and experience and developing consensus. It can be ignited by an event, commence with little or no structure, and yet deliver a result for collective benefit. When a tsunami struck the Southeast Asian region in 2004 with devastating consequences, many individuals and groups felt paralyzed about what they could do to assist the relief efforts in such a remote part of the world. While working as an IBM executive in Sydney, Australia, at the time, an employee of the division I was managing approached me directly to tell me that she had a connection with a local disaster relief organization that was preparing to charter a cargo plane to deliver much-needed clothing to victims of the disaster in Malaysia and Thailand. As the chair of the location's employee committe at the facility, which numbered around three thousand employees at the time, I rallied the team to quickly develop an e-mail, web, and text message communication campaign that was delivered to all employees encouraging them to leave unwanted clothing at the office dock for immediate distribution by the charitable organization to the disaster-affected areas. Without prompting, some employees forwarded the e-mails and text messages to family and friend connections to seek further support for the appeal. Other employees took it upon themselves to canvass their neighbors for spare secondhand clothing. Within four days of launching the communications, almost two tons of adult and children's clothing had been amassed by the staff. It would not have been logistically possible to achieve such a result had communal collaboration through connected groups and individuals not occurred. It's interesting to note that technology companies such as IBM, Microsoft, and Google have been developing products over the past fifteen years that automates collaboration within internal and external stakeholders.

Once the group has a reason to exist, the connected group is defined by the relative circumstance or their place in the world as it relates to their collective interest. Some good examples come from various religious groups. In recent times, the Catholic Church has been experiencing an exodus of practicing members from its ranks. In July 2008, a regular event on the Church's global calendar called World Youth Day was identified as the event that would not only unite the world's youth but re-energize the congregations. The event would be held in Sydney, Australia, and the star attraction was the inaugural visit of Pope Benedict XVI.

The Church launched a sustained communication campaign in the months leading up to the event. Web sites and blogs in numerous languages were developed and regularly updated to promote discussion and physical participation at the event. An online social network—xt3.com—was established for visitors or pilgrims, as they became known, to register to receive updates and interact with others in an online manner similar to the capabilities of Facebook. For the faithful on Facebook specifically, the pope updated his profile page; you could read his thoughts and ideas on various discussion forums and you could sign up for papal prayers delivered daily via text messages straight to your mobile device. There were audio podcasts developed and YouTube videos, all broadcasting the clear message of the propagation of the faith. The Catholic archbishops of Australia even commissioned a program of welcome for inactive Catholics called "Reconnect." The event proved to be "successful" in the eyes of the Church in providing a relevant context that Catholics who attended the event could relate to. And other organizations followed. In October 2008, the Australian *Sun-Herald* reported that the Anglican Church in Australia was planning a year-long million dollar initiative to "make Jesus known" called "Connect09."

Another example of the importance of relativity for the connected group is the Olympic family. The Olympics is an event that has stood the test of time, spanning three centuries. Its sustained longevity is a direct result of connecting interest in sport and the values of global unity and marketing it to its stakeholders by leveraging technology. It is challenged by having to manage subgroup interests, which the family embrace in the form of a group known as the Olympic Movement. The conglomerate of stakeholders, whose identities are published for the entire world to see on their web site, includes the International Olympic Committee (IOC), the Organizing Committees for the Olympic Games (OCOGs), the International Sports Federations (ISFs), the National Olympic Committees (NOCs), the national associations, the clubs, and the athletes. Their stated mission is to *"act as a catalyst for collaboration between all members of the Olympic family ...shepherd(ing) success through a wide range of programs and projects which bring the Olympic values to life."* The addition of broadcast partners, spectators, and its worldwide audience complete the Olympic family.

Every four years—and in the years in between—the Olympic coverage is enhanced by interactive web sites and games, streaming video broadcasts of notable events, and star athlete fan pages on social networks, such as Michael Phelps' Facebook fan page. The Olympic family is sustained through the organization's ideals that each stakeholder of this large group can relate to. A connected group's reason for being is to obtain a collective benefit from its very existence. Like connected individuals, these reasons may vary based on the group's motivation. It could be for personal fulfillment such as the case of tsunami relief donations, the spiritual belonging of World Youth Day, or a sense of belonging to the global village as is implied by the Olympic family.

* * *

The connected group has certain characteristics that set it apart from a traditional group. Those characteristics can best be summarized by the acronym *GROUP*. The connected group, due to its ability to connect instantly, regularly, and work collaboratively, can *gather* with great ease and efficiency. Due to their inherent connectedness, they have the capability to form expediently using familiar technology such as a web site or an online networking platform. When I was working at Australia's largest telecommunications company, Telstra, in 2008, I conducted an online social networking experiment on Facebook. A Telstra Fan Page group had previously been set up on Facebook with only thirty-four fans. The experiment was set up to test if I could increase the number of fans in seven days just by utilizing my Facebook friend network, which numbered around two hundred at the time.

The theory was that, based on a personal recommendation to join the group via an electronic invitation, I could get selected members of my network—whether they were Telstra employees or customers—to sign up as "fans" on the page. The goal was to achieve a target of one hundred registered fans in seven days by simply inviting contacts, effectively tripling the current base. The two hundred invitations were sent on Saturday, April 19. By the following Monday, the numbers had doubled to sixty-nine fans and by Thursday, tripled to 102 registered fans. By midnight on Friday, April 25, there were 116 fans registered—an increase of 341 percent during the test period. The test revealed both the power of personal recommendation and the ability to "gather" people together with a common interest, quickly and efficiently.

The other facet of a connected group is the effectiveness of its ability to *regulate* the group activities of its constituency. The formation of the family wedding event group was relevant only to the members who wanted to share and view the photos using the connection technology of an on-line network. The owner/manager of the group had the authority to moderate group membership and review input for relevance. This moderator could instantly remove and block undesirable members and edit out inappropriate comments and photos posted by members. The diligence of this "leader" in setting the tone for how the group functioned set the tone for the efficient running of the group. The connected group also had the capability to *organize* efficiently. Once again, due to the ability of connection technology to facilitate the formation of the group, they could get organized quickly.

The Telstra Facebook Fan Page was a standard group template provided by Facebook. The person who set up the page previously was automatically the group's manager (owner), controlled group membership, and moderated the discussion on group forums. The fans (participants) knew who they could interact with through the visibility of other members and what information and discussion forums they could view, and how to post items and make comments. A connected group is also *united* under the common interest of the group, within the context of relevance and circumstance. The members of the Telstra Facebook Fan Page were not members of Telstra's other competitors' fan pages and maintained their social loyalty to the group during the time of its formation. However, as circumstances changed, so did their "unitedness." If they chose to leave the group, they could easily remove themselves and, in one action, publicly signal their decision as other members of the group were sent a notification of the said member's departure.

When I left Telstra, the Facebook Fan Page was no longer relevant to my context and I felt no compelling loyalty to stay united with the group, so I left the group. Within minutes of my publicly visible online exit, I was inundated with messages from various group members and friends asking the what/when/how of why I chose to leave. A single action of departure created a sense of disunity that was immediately queried due to the visibility of the separation.

Finally, there's *participation*. The structure and nature of connection technology is to provide a platform for interactive individual and group communication. A connected group takes full advantage of this characteristic depending on member connection profiles and passion for continued interaction on the subject. Once the group fails to interact regularly, it invariably dies out. Like the connected group of participants in an audience wave, the connected group gathers in a single meeting site for a specific reason (in this case, a sporting event), comprehends the relevance of the planned activity to their state (boredom), organize themselves instantly (understands their role), unites in the desired practice and sequence (how they fit in the grander scheme), and actively and willingly participates (joins in). Once the activity is completed to the satisfaction of the participants, it concludes.

* * *

The key implications for the power of the connected group is in managing its presence for the greater good and harnessing its potential power for identifying opportunities. Like traditional groups, the connected group's influence and power are capable of being abused and utilized for extreme purposes. The good news is that all the information posted on the internet is usually "captured." So the messages, documents, and communiqués of connected groups can be monitored and, more importantly, traced back to sources by what I call "Webcrumbs." Webcrumbs are the trails of online evidence left by users who regularly upload, download, and post information on the web that gives investigators the ability to trace their activities. On the downside, some connection technology is also a potential threat to society. Lieutenant Colonel Timothy Thomas, in his paper on cyber planning, stated that "*The internet provides terrorists with anonymity, command and control resources, and a host of other measures to coordinate and integrate attack options.*" To manage this, governments of the world must act decisively in developing processes and policies that prevent connected groups from operating outside the realms of the best interests of society. In addition, individuals must be vigilant about their role in reporting suspicious group actions.

The connected group is also capable of identifying opportunities for business. Once connected, individuals passionately form a formal or informal connected group, and they can be incited to both interact and act. The tsunami relief effort at IBM is a case study that can be replicated for any worthy cause that requires an immediate response. There are also opportunities that benefit business, such as the case of popular American singer Justin Timberlake. As of November 2008, any Facebook member could search and access the following information about this famous celebrity: Justin Timberlake had just over 591,000 fans on his Facebook fan page. On this page were links to his official web site, thirteen music videos posted from YouTube, and the ability to download either his album from Amazon.com, individual songs from Apple's iTunes site, and assorted mobile phone ring tones.

His fans had posted ninety-four discussions about Justin, ranging from opinions about his music to which clothes suited the performer. In fact, the connected fans posted 7,523 comments and uploaded 1,235 photos onto the page. If you were a vendor of any product or service relating to Mr. Timberlake or his target market, you now had direct access to over 590,000 of his most ardent fans by name and their eyes and ears through discussion forums. This is a classic captive market that is defined by stated preference. If smart marketers were able to "befriend" these fans in a nonthreatening or nonintimidating fashion, there would be an opportunity to further clarify other preferences that could lead to the marketing of related products and services extensions.

As more individuals join online communities of interest that become connected groups, these groups gather and interact on various connection technologies such as blogs and online networks. These online networks provide a once-in-a-generation opportunity for society and business to be leveraged for both personal and commercial benefits.

CHAPTER ELEVEN
LEVERAGING NETWORKS

The internet is about people connecting to people, whether for business, politics, or socializing. That's something we've all been doing since long before the internet existed. The real accomplishment is to make those connections so versatile and different that they create a social network that not only reflects your life but expands it.
— Craig Newmark, founder of Craigslist.org, 2008

* * *

We're familiar with the old cliché: "It's not what you know, it's who you know." What's new about networking is that it's just as important **how** you network as whether you network. The *Connection Generation* collaborates by using various connection technologies. This interconnected system of people and devices may manifest itself through applications known as "online networks." The makeup of these networks is sometimes "invisible," as people keep their contacts and groups private from the outside world. However, connection technology by its very design also provides the capability and capacity for these connections to be publicly revealed, allowing people and devices to become "visible." As online networking provides the ability to identify, develop, and manage social and business connections via the web, a key benefit of participating in online networking is the advantage of access to millions of people globally who are a few clicks away from receiving an electronic message or invitation from anyone within the network. Online networking tools include the networks themselves and a range of collaboration applications to assist in maintaining those networks.

By now, most people have either heard of, been invited, signed up, or are passively or actively using an online networking site such as MySpace, Facebook, LinkedIn, Plaxo, and Ecademy platforms. These are generic online networks designed for social or business interactions between individuals and groups on the web. In addition, there are a variety of selective networks designed for specific interest groups such as the Catholic Church's World Youth Day site xt3.com, social sports-oriented networks such as Australia's 3eep.com and Affluence.org, which describes itself as *"an exclusive community of affluent people dedicated to making life better for both themselves and others."*

In September 2008, even the U.S. government launched a highly restricted online network to encourage members of the FBI, CIA, and other selective intelligence agencies to collaborate on a site called A-Space, which *Time* magazine dubbed "Facebook for Spies." Most people wonder how to benefit from spending time on these online tools beyond just finding old friends, messaging on walls, posting photos, and occasionally poking people. The main benefit of participating in online networking is that you can take advantage of access that is unavailable with traditional, real-world networking.

In September 2008, market intelligence firm Synovate conducted a survey of 13,000 people eighteen to sixty-five years old across seventeen countries. It reported that the global rate of social network adoption was 26 percent (one in four people), with the Netherlands (49 percent), United Arab Emirates (46 percent), Canada (44 percent), and the United States (40 percent) the top four nations. The survey noted that 37 percent of respondents from the United Arab Emirates said "they had more friends online than in the real world."

As an avid and active user of both LinkedIn for business and Facebook for social reasons, I've developed a methodology to best leverage these effective connection technology tools that I call the *Four Ps of online networking: Purpose, Profile, Participation, and Persistence.*

Purpose. Before you sign up to any of these tools, decide up front what you want to achieve, how you want to use these tools, and specifically set some goals. I started using LinkedIn and Facebook as contact databases for family, friends, and colleagues. Once I'd established my connections base, I decided to expand my network by reconnecting with extended family, friends, and former colleagues. The information and ideas I shared were of mutual interest, but I decided that I wanted to broaden my connections beyond people I knew to those who had the same interests around the world that I had not met. To achieve this, I decided to become an open networker. An open networker is someone who receives invitations to connect from people they don't know and, upon reviewing their profiles and interests, usually accepts their invitations. I began introducing myself to people with interesting profiles around Australia and in other countries and identifying groups with shared interests. As it works both ways, it also meant accepting invitations from people who I didn't directly know but whose profiles were consistent with my shared interests. I found the interaction and sharing of information, experiences, ideas, and concepts exhilarating, especially simply communicating with the first humans I had ever "met" from Bangladesh, Iraq, and Zambia.

The internet has been described as being either a time saver or a time waster. Clearly understanding what you want to get out of it determines your focus and how much time you spend on online networking. Once you settle on two or three purposeful goals, it should not only be a productive exercise but personally fulfilling.

Profile. In a highly competitive job and business marketplace, there exists a need for individuals and businesses to present themselves so either they or their services can be located instantly. Just as people and businesses choose to list themselves on White or Yellow Pages directories so their contact details can be found by others, creating an online profile is important to any person or business wanting to establish themselves in their target career or market. I often ask ambitious graduates and struggling small business operators if they've done a simple test—"Google" their names. If no listings are found, they are literally invisible in the online world to ever-present prospects, customers, and recruiters.

The reverse magic of online networking is that it can turn the invisible to the visible. Once you sign up to an online network, you're on the way to creating your own personal brand on the web. Search engines such as Google and Yahoo are optimized to source listings from global web sites such as global online networks. A profile on the business network LinkedIn is a unique opportunity to present yourself online, in a method that communicates the value proposition of the subject—individual or business—by the person who knows them best—you. By building and posting a compelling profile, you provide others with the ability to find you and attract other users who want to know you, potentially utilize your skills, sell you something you might value, just establish a connection, or even offer you a job. The time taken to develop a compelling profile that best represents you, including a photo, your key achievements, and your competencies, is an investment in you. On business online networks such as LinkedIn where more than 10 percent of users are recruiters, your profile is your online resume. A word of warning—due to the nature of the medium, what goes on the web stays on the web. So don't post items that you wouldn't want the whole world to have access to or know about. Remember the Webcrumbs. A good rule of thumb is to leave out the stuff you don't want your mother to ever know about.

Participation. I regularly hear comments from users saying that they signed up initially, didn't get anything out of it—so they left. In my view, this is like going to a party, sitting in a corner all night, and then complaining how bad the party was. Once you've got your profile up, be a participant—not a wallflower. The same social principles as a party apply—introduce yourself to others, start a conversation, share your thoughts, join a group discussion, ask questions, or play a silly game. Do something that shows that you're actively involved in the network and that you want to be there. In addition, choose to make a purposeful interaction regularly.

I've already mentioned how I use birthday reminders as an opportunity to keep in touch with my connections by sending a short note or messaging them with a birthday greeting. Another regular participatory action is to read your connections status updates. Most online networking sites provide users with the capability to update their current status in a few words by completing a status update. Users usually make comments about day-to-day thoughts, feelings, and special events. If I notice that one of my connections just passed an exam or got promoted, a short personal congratulatory message that takes fifteen seconds to type up and send is genuinely appreciated and goes a long way to better connecting as human beings.

Persistence. I recommend the discipline of setting time aside—some, not so much that it consumes you—to regularly participate in network activities and keep at it. There may be times when you don't get a timely response to an invitation, posting, or some other message. My advice is to persist. Some people may not be at the same level of adoption as you and are learning and using the tool at the same time. So be patient, stay on, don't give up, and eventually it'll pay off. I once invited a close personal connection to join my LinkedIn network. I worked with and spoke to this individual almost daily. Despite constant reminders, my invitation was ignored with no response. Eventually, when he moved on to another role outside the company, he forgot about my invitation and invited me to join his network. It was eighteen months after my original invitation but my persistence paid off as we eventually achieved the desired action. He is now the Australian CEO of a Fortune 500 company.

Like anything worth pursuing, to get the most out of online networking, you need to work on it. Make the time to clearly articulate your purpose and goals, get into the discipline of refining and enhancing your profile, actively participate in network initiatives that support your goals, and most of all—keep at it.

* * *

Once you've implemented the four Ps of online networking, a disciplined and persistent process is required to maintain your connection to the online network you've built to best leverage its potential. They consist of *relationships, research, and recommendations.*

Relationships. It's important to reflect on the purpose of why you chose to develop a network. Whether it was to retain a connection to existing contacts or to acquire a new set of potential friends or prospects, it's really about managing your "relationship capital." This can best be described as making old contacts in your address book come alive to become dynamic connections. An online network facilitates this relationship discipline by maintaining the currency of the contact. Any business expert will tell you that networking is one of the best ways to better connect with customers, prospects, and partners while advancing your career. For example, a business card or current contact address vehicle is reliant on that individual communicating any updates and modifications to that information as it happens. By their very nature, online networking databases are both dynamic and self-regulating. If contacts change their details or even jobs, they can advise their network all at once by changing their profile information. If they forget to change the information, the self-regulation of the network usually comes into play as their close connections send a reminder to the individual that their details haven't been modified on their online profile. Ultimately, these users update their profile details.

Research. To maximize the asset that is your network, it's important to use the database of your own or other networks for research. The best example of this is in job search. A majority of LinkedIn users are in the recruitment business. The system's capability to conduct advanced information searches on people and companies through various categories such as location, industry, and area of expertise makes it an invaluable resource to this community.

This same capability is open to sales and marketing personnel who can find out more about their existing customers through a simple inquiry by name and search new prospects through basic profiling by industry and location. Remember the example of Mr. Timberlake. You don't even have to know him to know the names of his biggest and most connected fans. It also pays to research online networking. Increasingly, more information about online networking is becoming available in book and blog form. A recommended resource is the business podcast called *The Connections Show*, hosted by Stan Relihan, which is available via download on iTunes. It captures the essence of doing business via online networking using interviews with experienced practitioners as case studies for business benefit.

Recommendations. In my experience, people and businesses appreciate being recognized for their value and contribution. It's said that it's always good to "pay something forward," as you never know when your time of need will arise. Business online network site LinkedIn provides a capability for people to write a formal referral for others that is visible to their connections. I recommend that you take the time to respond to requests for recommendations and even proactively write testimonials for those you feel rightly deserve an online "pat on the back" for their achievements. This may help in creating opportunities for others while establishing a visible reputation for you as a positive contributor to the online network.

* * *

Business networking is defined as the means by which individuals and groups connect for the common purpose of conducting business. Focusing on building a network and managing your connections is best summarized by senior recruiters when they advise that successful business professionals network when they have a job, not just when they need one.

Online networks provide business professionals with an unmatched flexibility to enable participants to make connections, share information, and post inquiries—at any time, from any place, across the country and across the world. It is a business application that allows businesses to get—and stay—connected to their prospects, customers, and partners. So why online networking for business? As a business, success and survival is reliant on promoting your business, meeting prospects, making contacts, building relationships, generating opportunities, nurturing centres of influence, joining communities of interest, and closing sales. Just as prospecting, relationship and brand building, and networking takes time in the real world, it's the same online—only more effective and efficient. There are three basic focus areas within online networking that, if used effectively, gives businesses a competitive advantage to effectively and efficiently manage their network of stakeholders. They are: people, groups, and expertise.

People. By November 2008, LinkedIn boasted thirty million business professionals as members internationally with representation from executives from all Fortune 500 companies and, according to LinkedIn, 46 percent were business decision makers. Having the ability to search this dynamic database of business professionals has obvious benefits for sales and marketing personnel in any business. In addition, once a customer, partner, or prospect has accepted an invitation to join your network, you establish a visible online connection with that person, allowing the potential to better relate, communicate, and exchange ideas, information, and opportunities—at a one-on-one level.

One of the inherent values of online networking, both social and business, is that you don't have to "know" someone directly to reach them. The network, by design, provides the function of reaching out to someone through a direct or indirect connection, through an interest group they are members of, or self-profiling. It's not a replacement for physical or tactile interactions—it's an enabler to develop deeper relationships. I've presented to sales people on the steps required to search information about their key executive contacts that they're about to meet using tools such as LinkedIn and Google. Invariably, they automatically see the value of gathering even the smallest pieces of information to establish rapport and in building better relationships.

My favorite example is of a sales representative who was about to invite his CEO customer to the ballet as part of a corporate hosting initiative. A quick search on a social network showed the customer profiling himself as an ardent football person who coached his son's team, was president of the local club, played in the over-thirty-five team, and was a big fan of the state's team. The ballet invitation changed to a box seat for the customer and his son to a major game of his state's football team. The effort required by the sales representative to find this information was minimal and the value to the ongoing customer relationship was priceless. People shouldn't waste time communicating to their target audience unless the most basic individual research is conducted on them through Google or an online network. You can find out a lot about people not just directly, but from people who may know, work with, or are associated via a group to that individual.

Groups. An important feature of tools such as LinkedIn is to search for, join, and even create groups with shared interests. By joining these groups, you can virtually connect to other group members, join discussion sessions, and share information and ideas. The ability to harness the different perspectives, experiences, and skill sets of this collective in an instant and cost-effective way must be an essential resource to tap into for any business.

LinkedIn contains the most diverse groups. There are profession-based ones such as the Sales and Marketing Community, Linked HR, Linked Lawyers, and the Top Executives Network (TEN); country-based such as Friends of India, Australia Network, and the American Small Business Coalition; and initiative-based such as Online Lead Generation, Think Green, and Women for Women International. The online networks themselves are a potent platform for realizing direct commercial opportunities.

In a press release posted on their web site on October 15, 2008, global pizza chain Pizza Hut announced a new application that offered the ability to order its pizzas—on Facebook. Previously, Pizza Hut had introduced a number of ways to order its products by using connection technologies such as the mobile web, text message, or from their web page. *"We are moving fast to put our online customers in charge—any way they want to order from us, we'll be there for them,"* said Bob Kraut, VP, Marketing Communications. *"Facebook is the next logical step for Pizza Hut. As Facebook's popularity grows among younger socially connected consumers we have the opportunity to provide a pizza experience they will love—our convenient ordering combined with Facebook's relevance."*

Expertise. Establishing a broad network enables you to turn to different experts or groups, depending on your professional challenges. As mentioned in a previous chapter, in February 2008, Bill Gates joined LinkedIn and, shortly after, posted a question regarding how to encourage young people to pursue information technology and science careers. The question was posted on the site's Answers application, which allows a user on the network to ask a question that can be answered by any other user, which was potentially twenty-five million business professionals at the time. This is a classic case of the "wisdom of crowds" application. More contemporarily, it's the global equivalent of the ask the audience option on the TV game show *Who Wants to be a Millionaire*, where the contestant can ask the opinion of the gathered audience to a question being asked and receive the most popular answer. Within two months, Gates received 3,566 responses, which his army of staff could wade through to find the gold nuggets that could be pushed through various stages of implementation. The cost and efficiency benefits of conducting this kind of research are incalculably valuable.

In 2008, the senior marketing manager of a Fortune 500 company told me that she researched and surveyed her target market using LinkedIn to develop her strategic marketing plan. In the context of living in an increasingly connected world, online networking platforms are not only additive but essential platforms to successful personal and commercial connections. In today's business world, it's more than what and whom you know—it's how connected you are to people, groups, and experts in the exchange of information and ideas for commercial value.

* * *

In November 2008, after being on LinkedIn for almost fifteen months and implementing the discipline of managing my profile and connections, I accumulated more than 3,600 connections. At the time, I also had over four hundred connections on Facebook. I have regularly been asked the question of how to manage this volume of connections' questions without spending a lot of time and effort to do so. As mentioned, my first experience with these networks was to do what most people do and invite only people I knew. In fact, that's the guidance given by most online networks. However, my perspective changed when I listened to the perspectives of one of the Top 20 most LinkedIn users in the world, Christian Mayaud, on an episode of *The Connections Show* podcast.

Christian was discussing high-performance systems management for technology networks. He related his experience on the best method for efficiently managing and forecasting the heavy traffic on a technical network to get the best performance out of the system. Firstly, there is reviewing previous network throughput, which he called Former

Active Network or FAN. Then, a review of current traffic is conducted, called Current Active Network or CAN. Finally, there's the projected traffic which is called the Potential Active Network or PAN. Christian concluded that managing a large online network of human connections effectively required a similar process. An individual could manage his or her FAN (former school friends and work colleagues), which would number around 150 people, and CAN (current family, friends, and work colleagues), which would be another 150 connections. This sum number of 300 is the average contacts on a business professional's mobile phone contact list. The individual's PAN is limited only to the purpose of having an online network. As an open networker who intended to broaden my network to maximize the amount of information, experiences, and ideas available to me, I figured that it was better to be connected to someone now, in case I needed to assist or call upon that prospect in the future for whatever personal or commercial opportunity would arise.

I'm also asked about which online networks should one join and how many is too many. I like analogies and, for me, although the phrase "online communities" is used to describe these sites, I think they are more like parties or events. I've regularly attended a number of business events such as business functions, corporate event after-parties and corporate client hosting. I also attend personal and family social functions like birthday parties, weddings, and christenings. At these events, I meet various people—some of whom I "connect" with either commercially or socially, and some of whom I don't. The analogy for me is that online networks are huge online "functions." I "attend" LinkedIn, Facebook and Twitter parties more often, as there are more people I connect to there than others. It doesn't mean I don't "drop-in" to the other parties such as Plaxo and MySpace from time to time to see what's going on. My advice as a connected individual is to experience as many online functions as you can manage, but participate and persist with the ones that regularly fit your stated purpose.

While there is a potential for abuse, it is important to fully comprehend the positive power of groups and networks. These communities have both social and market powers, which, if harnessed for good, have the capability of generating ideas and opportunities to change society and business as we know it. Conversely, they can be cultivated for negative intentions that could result in adverse impacts to the world.

CHAPTER TWELVE
MANAGING CONNECT GEN

Only through our connectedness to others can we really know and enhance the self. And only through working on the self can we begin to enhance our connectedness to others.
– Harriet Goldhor Lerner

* * *

During one of my management roles in corporate online sales, I interviewed a young graduate, Julie, for a junior sales role on the team. She worked on a sales team but had no direct online sales experience. I asked her why she wanted to work in online sales as opposed to traditional sales. "*Online is the future of business*," she declared. Her generational profile had her sitting squarely in Gen Y and although she was raised in an online environment, it intrigued me that she still thought it was a "future" technology. My next three questions were more about testing her thinking than assessing her abilities to perform the role. Firstly, I asked her when was the last time she wrote a letter to someone, to which she responded, *"Can't remember—I use e-mail now."* I then asked her when was the last time she went to the local library to do research. She said, *"Ages ago, when I was at university. If I need to find anything now, I Google it first."* My last question was directly sales related. Being in her twenties, I assumed she went to concerts with her friends. I asked her if she purchased tickets over the counter at the local box office prior to attending these events. Her response was *"Hell no, I buy them online like everybody else—you get better deals there."*

I came to realize that although Julie was Gen Y, she was a *passive* connection profile. I assumed that because she was more familiar with connection technology than those older than her, she appreciated its value in the business world. The reality was that although she knew how to connect online to people and information instinctively and it was as "normal" to her as running water and electricity, she didn't comprehend that the applications she was using were a connection technology of today rather than of tomorrow.

This is one of the challenges in managing the *Connection Generation*. *Connect Gen* is a collection of diverse connection profiles, groups, and networks from various generations and perspectives. The question arises as to how to best manage and leverage the mix of individual profiles and groups to extract the best value from them. To answer this question, it's important to understand the threats and opportunities that underpin each connection profile.

Basics possess an overwhelming denial of their ability to learn and adopt connection technology. This limits their propensity to consider technologies that may enable them to become more connected to the rest of society and business. This limitation has the potential for them to be "left behind" from online social communications and events that could lead to the real threat of this group not only being uninformed but of missing opportunities altogether that may benefit them. Due to their attitudinal unwillingness to consider connection technologies to better connect themselves, they could be considered social lepers in an increasingly connected world. Their reliance on connection agents to act on their behalf will be short-lived as these agents come to the realization that they have less time to spend on tutoring people who are perfectly capable of familiarizing themselves and utilizing connection technologies. It may ultimately test friendships and relationships.

On the opportunity side, *basics'* same dependence on the connected behavior of others may bring out their willingness to consider the practical benefits of the more applicable connection technologies as they relate to their day-to-day activities and relationships. They may even be "dragged into" the online world if specific connection applications become the standard method for interacting and sharing relevant information amongst their circles of influence. The pervasiveness of mobile device technology and its applications may be the vehicle for greater connectivity for *basics* and potentially "upgrade" themselves to more connected profiles.

Passives have their challenges too. *Passive* Julie the graduate relied on connection application to fulfill a specific need or process without realizing that it was not new media but a *NOW* medium. *Passives* have a tendency to underestimate—and even deprioritize—connection activities. Although they understand the technology and, like Julie, are proficient as to its use and application, they limit it to a single, logical application or process. Even if they've taken the plunge to adopt a particular technology, they may not persist with it if they don't see the value. As the old saying goes, you can lead a horse to water but you can't make it drink. If *passives* can't see the value of connectivity, they may be left behind socially and in business, potentially even missing out on job opportunities, as Julie ultimately did on this occasion. On the plus side, *passives* have a propensity to test and trial new applications and they understand the importance of staying connected to close family and friends. Like *basics*, if more of their social and business circles used connection technologies as standard modes of connection operation, they would be more inclined to try them out and increase the rate of permanent adoption.

By design, *selectives* prefer to keep "closed" networks of connections. They're very specific about who's in and out of their social circles, which limits their ability to obtain and understand other points-of-view on issues and challenges. *Selective* business professionals would have very deep relationships with existing customers but would struggle with prospecting for new business. This perspective, if not managed by their inherent disciplined approach to connecting, could see them regress to a less connected profile. The opportunity for *selectives* is that although their networks are closed, their minds are open to new technologies and ideas. They link the logic between application and technology quickly and have a tendency to be more productive due to their focus on specific connection activities.

In the case of *actives*, management is required to ensure they become productive contributors. They tend to drive connection activity at alarming speeds without stopping to assess the quality of their efforts, leading to unproductive time and activity. It's essential that a management system that incorporates specific controls be implemented by managers of this group to ensure focus is directed in the optimal areas of connection. They are also impatient about job situations that don't take advantage of their connection abilities, which could potentially create a disincentive and reduce motivation. As mentioned previously, the opportunity for *actives* is that they make for great sales and marketing personnel. Their capability to initiate interactions and proactively connect for commercial benefit should be leveraged by any business.

For *supers*, one of the threats is that they tend to be intrusive and distracted easily. Their addiction to connecting to anything and everything may seem intrusive and offensive to other profiles, which should be managed via their own disciplined system. They can be distracted by the thrill of hunting down multiple contacts without reviewing each connection's specific value to its stated purpose. This could also lead to a loss of productivity and commercial output if left unmanaged. The opportunity is that *supers* can act as connection mentors to other profiles to boost effective and efficient connectivity for value. They have an intuitive and instinctive sense of how to best leverage connections and their high degree of expertise and experience should be openly shared.

* * *

With the mix of diverse profiles, there are guidelines recommended for how to manage individual profiles for personal and commercial benefit. I summarize these four factors as *personalize, listen, activate, and nurture*, or more easily remembered as *PLAN*.

Personalize. Each of the individual profiles, regardless of their degree of connectedness, relishes the importance of a personalized connection to others. *Basic* Geoff the taxi driver understood the importance in his business of staying connected through technology by his adoption of mobile technologies to keep him connected to requests for his services from customers via multiple connection mediums. He recognized the link between connection technology and personalization by describing his wireless mobile earpiece as "*my Bluetooth.*" *Passive* Pooja the professional needed to personalize her online business profile to become more visible and ultimately more competitive in the marketplace she chose to target. *Selective* Richard the sales manager developed deep personal relationships by utilizing his online networks, both socially and for business. *Active* Rachel the student also used her social network as a way to personalize communication with her friends Chris and Kim to avoid potentially embarrassing gaffes about the status of their on/off relationship. *Super* Mark the concierge translated his online connectedness proficiency to meet his job goals of ensuring positive guest interactions with the hotel. Understanding the role of personalization is an important factor in better managing connected individuals.

Listen. Listening to what individual profiles are doing, commenting, and providing feedback is essential for managing connection profiles. A case in point is in identifying the most relevant connection preference for effective connection management. If you were to manage the following individuals effectively, you would need to find out and understand their preferences. By determining connection profiles and analyzing their behavior, we realize that *basic* Damian the plumber prefers text messaging with friends, *passive* Diana the executive relies on her personal assistant to stay connected, *selective* Kerry the student uses instant messaging to link her with other students, *active* Maurice the retiree uses e-mail with his contacts, and *super* Ron the recruiter loves to connect on LinkedIn with just about anyone.

To capture the hearts and minds of these individuals, it's important to activate the radar to ascertain how they connect. In the case of *passive* Diana the executive, her choice of connection vehicle happens to be another human being, *active* Seamus the assistant, who is her indirect access to connection technology. If you wanted to relate to her and get her attention, you would not go directly to her. You would have to start by contacting her assistant in the first instance and asking him to build a connection bridge detailing the reason, relationship required, and relative return (*three Rs of connection*) for her to want to connect. By utilizing an identified connection preference like an assistant permits a relevant method of getting and potentially staying connected.

Activate. For some connection profiles, it is a challenge to establish a connection. To better manage these individuals, regardless of their capacity levels, it's important to activate an online connection. *Basic* John the senator connected to his office by having his e-mails printed out for him, *passive* Martha the homemaker tried to develop her own blog to promote user feedback, *selective* Bill the technologist tried Facebook before settling on a profile on LinkedIn, *active* Oprah the media mogul uses her web site and global television show to rally her dedicated fan base, and *super* Barack the president links to his constituents one voter at a time. The success or failure of each of these examples is the point of connection management. If not planned and executed well, it will fail the purpose and intent. Done well, it can not only win presidential elections, but potentially change the world.

Nurture. Once you've understood the importance of personalization, leveraged their connection preferences, and successfully activated connections with stakeholders, nurturing established relationships is key to managing these connections longer term. *Basic* Mary-Jane the teacher remains connected by friends despite her remoteness through regular phone calls to family and friends. *Passive* Loretta the manager continues to make outgoing mobile phone calls to the people she needs to stay in touch with. *Selective* Melissa the hairdresser complements her online network presence through her preferred face-to-face meetings with friends. *Active* Hamish & Andy the comedians continue to entertain their fan base through multiple opportunities to interact, and *super* Karl the college student maintains friendships he made in Australia for life through living his regular Facebook presence.

As simplistic as it may sound, the key steps to take to managing connected individuals to best realize their roles in a connected world is simply to work through the *PLAN*.

* * *

It's acknowledged that connected individuals are part of connected groups and networks. An additional set of initiatives is recommended to best manage them, which involve *opportunity, participation, engagement, and navigation,* or, in summary, *OPEN.*

Opportunity. This one's about understanding and working with the precise goal of the group or network and taking full advantage of its collective purpose. The Olympic movement has to sustainably galvanize diverse groups and individuals with different political and sporting interests, who speak different languages, for an event that occurs once every four years. Their web site presence that encourages inclusiveness and regular interaction is one of the anchors in realizing their Olympic ideals and values—and keeping the virtual Olympic flame burning.

Anyone can seize the opportunity and create a following if the benefits are clearly articulated. British web site storemob.com exists for the specific purpose of organizing groups of shoppers to turn up at the same store and at the same time to obtain group discounts. Their web site declares, *"Shoppers visiting the same store can haggle prices more effectively when working together than when acting alone. This is because the store will usually discount more if they sell more."*

According to the Australian *Sun-Herald*, the trend is also being adopted in China, through sites like TeamBuy and Taobabo that let subscribers *"find like-minded shoppers."* They state that Facebook is also being used as recruitment tool.

Participation. It's often difficult in a time-poor, easily distracted, and information-overloaded world to get people to be part of something unless they see the value. It's possible to transcend these potential barriers by establishing compelling propositions and communicating the rationale of being part of a greater group. The tsunami relief initiative at IBM was a question of appealing to the sense of community to assist those with a greater need by taking single actions whose sum would result in a grander collective effort. The Catholic Church's 2008 World Youth Day relied on people taking part in each of the connection activities to emphasize the event goal of promoting unity among the faithful.

Engagement. Once people are willing to play and have bought in, the connected group needs to be engaged to realize the fullness of the opportunity at hand. For example, there are multiple cases where text messaging has not only engaged but incited disgruntled groups to action. A *Newsweek* article in August 2006 commented that events like the mobilization of demonstrators in 2001 to rally for the ousting of Philippine President Joseph Estrada was a direct result of text messaging. A texting campaign resulted in the unprecedented youth voter turnout in Spain's 2004 presidential elections and, in 2006, Mexico's President Felipe Calderón sent millions of text messages days before his narrow victory over his opponent. Why this potency of text messaging over other mediums? Maybe it's because it's on a device that's carried around by the person, instantly accessible, is cost effective, and gives the individual a choice to engage with or not. The *Newsweek* article mentions that the chances that people will open and read a political text message is 95 percent versus a less efficient 5 percent to 25 percent for e-mails.

Network. The power of connecting individuals and groups together to form networks should never be underestimated, both for good and for evil. These networks can be created in a very short space of time. A small "network" like the Telstra Fan Page tripled in size in a week. In the years between the 2004 Athens Olympics and the 2008 Beijing games, the online networking site MySpace was launched and took less than two years to become the No. 1 social networking site in the United States with over one hundred million members. If it were a country, it would be in the top ten in the world. It made News Corporation's Rupert Murdoch a quick and healthy profit when he purchased its owners Intermix Media for $580 million in July 2005 and just over a year later secured a $900 million deal with Google for search and advertising capability on MySpace.

But it's not all for good. A seventeen-year-old American high school student, Allen Joplin, was shot dead by a gatecrasher at an underage party after it had been publicized through MySpace. In Australia, sixteen-year-old Corey Worthington received international notoriety for using online social networks to promote his party, which escalated into a riot. Networks are powerful and need to be planned and regulated to deliver benefit.

It's a challenge to manage groups to extract the best value out of them. Just ask any sporting coach in the world. With connected groups, a good understanding of the principles of *opportunity, participation, engagement,* and *networks* provides the framework for more effective management. In managing and dealing with them, it helps to keep an *OPEN* mindset.

When it comes to planning and managing expectations of human interaction, the question of the effectiveness of online interplay versus traditional physical—or offline—interaction is often raised. The reality is that it's not a question of one or the other—it's a question of preference management and applicability to the situation. Some connectors like *basics* and *passives* take a backseat when it comes to connection technology and have a preference for face-to-face or phone interaction. *Selectives* see online as a supplemental connection tool in their daily lives and use a mix of real and virtual interactions to stay connected while *actives* and *supers* live comfortably in both worlds.

The key here is not about online replacing offline. The most effective connection methodology is the one that works and is prefered by the parties concerned relative to what needs to be achieved. I call this concept *inline*, which means to establish the most relevant and effective connection methodology, which as the name suggests is "in line" with what needs to be done.

For example, most would agree that it's much more personal and sociable to be physically present at a family or friend's wedding celebration. However, what if two of your friends invited you to their more practical and virtual internet wedding? That was the case in the wedding of Andrew Hunt from Somerset, England, and Lisa Grosso from Florida in the United States, who, in 1996, became the first officially recognized virtual marriage ceremony over the web. They were married via Internet Relay Chat (IRC) by a minister in Seattle, Washington, and blessed by a vicar in Lyndhurst, England. It's doubtful that this celebration of their love and union would be any less special for the bride and groom than a face-to-face wedding gathering. Yet it was achieved *inline* with traditional practice and connection technology to meet the desired goals of the parties concerned.

Inline is a critical consideration for business, especially in the case of relationship management. While telephone and video conferencing is being widely used by organizations to save time and expense of international and interstate travel, there is no replacement for establishing and maintaining individual relationships and group dynamics via regular face-to-face social and business interaction. In my corporate experience, most of the value from team meetings, workshops, and conferences was achieved outside of the formal agenda sessions, usually during pre-event breakfast meetings, coffee breaks, and post-event drinks at the local bar. Some of these interactions have developed into friendships and associations that transcended corporate life and are being maintained through connection technologies like online networking, *inline* with the current and applicable circumstances.

The customer relationship perspective is another example. Many business decision makers are time poor and cannot always meet with their vendor representatives as regularly as required to maintain good business relations. While eager account executives would kill for regular customer face time, they sometimes have to settle for a phone call or e-mail. Savvy customers are using a connection technology called the personalized URL, or PURL. A PURL is a web page customized by vendor account managers to provide online relationship management between the sales team and one or more customers. It allows electronic interaction in a personalized manner, creating a regular connection for sales opportunities and leads. As a complementary medium to regular face-to-face meetings, it's an example of how a sales representative can better manage a connected business relationship *inline* with the customer's situation and expectations. It's also an increasingly important consideration for all businesses in managing the connected consumer.

CHAPTER THIRTEEN
CONNECTED CONSUMER

The findings surprised even us as today's digital consumers have moved well beyond merely sampling Web 2.0 technologies and services. They are now adopting these services at a breakneck pace and readily experimenting with new, more sophisticated offerings en masse.
– Razorfish commenting on FEED: The Razorfish Consumer Experience Report, October 2008

* * *

While a person's individual connection profile determines his or her attitude, behavior, and capability toward being connected and connecting, the person invariably belongs to a connected group or network that acts in a collective manner. The other key factor is the role the person plays in a specific context relative to society and business. To explain one of these real world roles in business, allow me to introduce Kira, part of an evolved breed of customers called the *connected consumer*. The connected consumer presents both challenges and opportunities to businesses that market, sell, and provide services to this group in an increasingly connected world. They don't just use technology to communicate. They "utilize" devices and applications to be kept informed, exposed, and connected to the people, information, and ideas of their choice and desire that suit their busy lifestyles. They possess inherently different consumer behaviors of those who lived before the *Connection Generation*.

Kira is an office manager for a boutique recruitment agency. She's just turned forty-two, has been married to the same man for nearly twenty-five years, and has two adult children approaching their twenties. Kira starts her Monday morning by waking up at 6:00 a.m. to the dulcet pre-programmed tune on her iPhone alarm. She figures that there's no point having a separate alarm clock as she can use the mobile device that she needs to have by her bed as she sleeps, in case she gets an urgent call from her elderly parents or her boss. Having the device bedside also serves as a reminder to charge the phone battery overnight anyway.

She gets herself ready by showering and applying the vast array of facial and body lotions purchased online for a fraction of what she would pay in a retail store. She sprays herself using the last bottle of authentic French perfume, which she bought directly from the supplier's global web site. They used to deliver internationally, but a ban on shipping alcohol-based products to Australia due to terrorist threats means it's the last bottle she'll own for a while. She puts on the dress she ordered online ten days ago, which was delivered last Friday. She found it in the monthly direct mail catalogue she subscribes to, which is sent to her home address.

She hops into the car with her husband for the trek into the town where she works. As they reach the usual traffic snarl on the main motorway, a distinctive beep is heard from the car's dashboard. She assumes that it comes from the fuel indicator, compelling her to reach for her iPhone to access the iGas application, which lists the latest fuel prices around the area she is traveling to. Her husband advises her that the alarm is in fact a reminder that the vehicle's 20,000 mile service is due.

The traffic is particularly heavy this morning so Kira pulls out her iPod where she had previously downloaded a series of ilectures for the bachelor's degree course she is completing part-time at the local university. She knows she can download and listen to the lectures on her iPhone but she hasn't had the time to switch from the more familiar iPod device yet.

They arrive at the usual café where they have a quick breakfast. During breakfast, she mentally recalls the situation where she had to send an overdue university assignment to her course leader, to which her solution was to use her laptop to access the café's wireless network. Before leaving the café, she synchronizes her day with her husband—she on the iPhone and he on his BlackBerry. He informs her that he has one meeting to attend this morning and will be working from home for the rest of the day.

After picking up a newspaper on the way in, she walks into the office, drops the newspaper on her boss's desk and powers on her office desktop computer. While the system is booting up, she listens for any voice mails left on the office answering machine late Friday afternoon or over the weekend. When her system is ready, she logs on to her e-mail, instant messenger, and reader feed applications. She looks through her e-mail inbox and prints out all the important ones. She deletes all the junk mail not picked up by the computer's anti-spam software. She also reviews her reader feed for news items of the day, which are fed directly from her chosen predefined global news web sites. She has no need to read a newspaper.

Her boss arrives and thanks her for coordinating last week's installation of his office cable TV, as he prefers to watch news and stock reports in his office. She hunted down the best cable deal on the web by comparing various packages and arranged for the installation contractors to be available during times that suited their business, with which they obliged. The previous week, she saved the company $500 in office copy paper costs by tracking down the best promotional offers, including delivery within four hours, again on the web.

Her first action item of the day was to register to attend a weekend course in first aid. As the office didn't have a trained first aid officer on the floor, she had to complete a certificate course conducted by the local Red Cross branch. She completed the registration process online within ten minutes, entering the date, time, and venue details into her electronic diary. She then accessed the online company bank account transactions for the previous week, imported the data onto a spreadsheet, and commenced reconciling the accounts. She also manages to check her own credit card balance while she's on the bank's web site.

During the course of the morning, an instant message pops up on her screen from a close friend inviting her to lunch in the city next week, which she accepts—and schedules in her e-diary. The lunch invitation is at a new Italian restaurant that opened to rave reviews a month ago.

Just before lunch, she takes a walk across to the local post office to collect the company mail. She notices that the temperature outside has dropped and remembers that she doesn't have her coat with her. She calls her husband on her iPhone and asks him whether he can bring one of her jackets with him when he picks her up later that day. He innocently asks, *"which one?"* to which she replies, *"the cream one."* He looks in the closet to find at least three cream-colored jackets. After some discussion about its precise description, he says he'll have a better look and vows to get back to her. He picks out the one that closest fits her verbal description, lays it on the bed, grabs his BlackBerry and uses the phone camera to take a photo of it. He then sends her a video text message, attaches the photo with the message *"this one?"* to which she replies, *"yes, dear, thanks."*

She is impressed with her husband's application of the technology and makes a note to speak to her boss about how they can use this video text messaging capability to present employment candidates to their time-poor yet demanding clients instantly via their mobile devices. As she's busy this particular day, she settles for a sandwich at her desk for lunch. She accesses her Facebook home page to find out what her friends have been up to and uploads selected photos from the birthday party she attended on the weekend. Her page also reminds her that her nephew's birthday is in two days' time. She accesses a local online variety store and arranges the purchase and delivery of a Transformer toy suitable for an eight-year-old, with a personalized birthday card included. She then receives an e-mail—which makes her do a happy dance in her head—advising her of her successful eBay bid for the designer bag she had her eye on. As lunchtime ends, she remembers to renew her monthly gym membership, which she completes online.

As her company happens to do business in New Zealand, she has to comply with regulatory requirements to lodge company returns, which can only be done online. She spends most of the afternoon preparing and finalizing the submission prior to final review by the company accountant. The day closes with the receipt of a text message reminder from her physiotherapist of her previously arranged appointment at 7:30 a.m. Tuesday morning (tomorrow) and another text message from her husband asking whether she wants him to bring her coat up to the office, which he does anyway. They join the traffic jam on the way home and her husband puts on a podcast recording of motivational speaker Tony Robbins's speech at the 2006 TED conference, which he downloaded from TED.com earlier.

She gets home and is relieved to remember that Monday night is pizza night. The kids organize for the pizzas to be delivered with the transaction completed on Facebook using Dad's credit card. The first human involved in the transaction is the Pizza Hut delivery person who arrives at the door with the pizza order. They dine as a family and chat about the events of the day. After dinner, Kira takes out her laptop and powers it up. All the home computers are linked to a wireless broadband setup. She accesses her university course web site and participates in a previously arranged online discussion with other students. She reviews the course bulletin board and finds that the latest ilecture is available, which she promptly downloads to her laptop and uploads to her iPod. She accesses her home e-mail account and is reminded that the electricity bill is due, which she immediately pays online with her credit card. She also receives this week's edition of her husband's blog post via e-mail, to which she is a regular subscriber.

She looks at her follow-up file and sends a confirmation reminder to the hotel in Christchurch, New Zealand, that she'd booked a few weeks back as part of the family holiday in December. Her physiotherapist recommended some good walking tours of the South Island and her Kiwi boss from Queenstown gave her some tips on the best places to stay based on his local knowledge. She researched, booked, and paid for the entire three-week vacation trip—including flights, transfers, and accommodation —online without speaking to a single travel agent. Her final task this evening is to visit Google Maps and access the street view feature, as she is curious about the restaurant where she will be lunching with her friend later in the week. Just before bed, she hooks up her iPhone to its charger, checks that the alarm is set for 6:00 a.m., and is reminded that her first appointment is with her physiotherapist at 7:30 a.m.

* * *

Not all consumers are like Kira—yet. As more and more people understand the power of connection technologies and how they can benefit them personally and commercially, more people will transform into connected individuals who instinctively use it as part of their daily lives. There are three important characteristics that every business that targets this group needs to be cognizant of to successfully acquire, retain, and relate to them.

Information Gatherer to Resource Hunter. First and foremost, connected consumers evolved from information gatherers to resource hunters as they interacted and experienced the various stages of the developing web. This began with the early "navigating" days using Netscape Navigator in 1995, further exploration using Microsoft's Internet Explorer in 1997, and a year later, using search and discover tools via Google and Yahoo. In the last five years of the twentieth century, they knew the information was out there and tried each tool to gather information relevant to their needs. This was a time-consuming exercise in the early days but served them well for future prospecting.

In the early twenty-first century, this group graduated to acquire the capability to go beyond browsing to being able to figure out exactly where in the World Wide Web they wanted to go, using a mouse, any internet browser, and a search engine. If Netscape showed them the window to the web world and Microsoft's Internet Explorer unlocked the front door, Google and Yahoo opened the floodgates to tracking down information and ideas. The combination of these elements turned the gatherer into a resource hunter. Once they had the knowledge of how to navigate, explore and search the web world, they soon developed a laser-focused efficiency about how to effectively and efficiently track down what they were looking for.

As a hunter uses guile, skill, and local knowledge to catch its prey, the connected consumer uses a combination of search proficiency and acquired experience. Kira discovered where to find her favorite clothes and perfume from various search missions. She bagged her designer accessory on eBay through various failed attempts at bidding, before learning what it took to capture and win the desired items. Her capability to contrast and compare the best cable TV or office copy paper deals for her business office was also a direct result of resource-hunting skill and experience. This included the capacity to negotiate "deals," which in both cases included the added services components of installation and delivery.

Exposure Efficiency, Application Relevance, and Easy Access. The second characteristic of the connected consumer is the three-pronged combination of exposure efficiency, application relevance, and easy access. Kira is selective about which device and application she uses for what purpose. She's learned that an iPhone can double as an alarm clock. E-mail is for information she needs, which means irrelevant messages are junked automatically or manually if need be. Reader feeds filter only relevant broadcasts—in her case, specifically chosen news and local weather sources—that she wants to expose herself to. She has no use for a newspaper that contains mostly advertisements and some general information that is not directly targeted to her interests. Connected consumers have developed a filter management system that detects information and messages they DON'T like as efficiently as it tracks things that attract and appeal. I call this exposure efficiency—the capability to choose to be exposed to devices, information, and ideas that make one more efficient and effective, which in Kira's busy life is an important characteristic to possess—and a godsend.

Application relevancy is the ability to equate an application to its specific purpose. The alarm beep from Kira's car creates a message in her brain that connects a sound to a prompt to put more fuel in her car—that triggers another action to access the iGas application on her iPhone to search for a cost-effective fuel solution to quench both the continued alarm beep and the thirst of her car's engine. Her thinking is programmed to know that an iPod, while usually associated with music, can also be a source of learning and knowledge by accessing downloaded lectures. She knows the iPhone can also incorporate this but her iPod will do for now, as it's what she is used to and is easily accessible.

Easy access is simply the ability to find information instantly through any information vehicle. Kira is capable of accessing the clothes merchant's catalogue via their web site. She chooses to use the hard copy catalogue due to its instant availability. She doesn't have to power or log on to her computer to view it. The document is physically delivered to her home address, which in itself serves as a prompt to review it. It resides by her bedside so she can pick it up whenever she is in the right mood to read it and, due to its portability, she can take it to the lounge room or outside to peruse its contents. Once she decides what she wants, she knows which applications to use to source and procure exactly what she desires. She also knows she can walk to the local Red Cross office, visit the bank, and shop at the local toy store to register for courses, access accounts, and shop for gifts, respectively. She is also savvy enough to know that she can complete all these tasks online—easily and efficiently—in at least one-tenth of the time it would take in the physical world and *inline* with how she personally operates.

Trust and Relationships Are King. The third characteristic is that trust and relationships rule. Connected consumers strongly value social friendships and contractual relationships in line with their needs and wants. They harness the knowledge of "experts" in their circle of influence, which includes family, friends, colleagues, and suppliers that have been developed and maintained through time. This knowledge network is often consulted for support, advice, guidance, and insights into where and how best to meet many requirements. Kira possesses an exclusive set of trusted family and friends who are granted direct access to her online world through instant messaging. Her friend is granted permission to interrupt her working day to invite her to lunch—a message she chooses to accept or reject. She is selective about letting people into this environment and wouldn't allow marketers or hawkers to enter this—or her private Facebook space. From a provider perspective, trust needs to be earned. Kira has a long-term relationship with her physiotherapist that permits the sending of text message appointment reminders. Her casual hairdresser is not given the same privilege due to the nature of their relationship.

In the case of the cable TV installation for her boss, she is adamant that the contractors arrive as arranged. From her point of view, if they had not arrived and completed the task as arranged, this could have resulted in irreparable damage to any future business relationship between her and the cable company. The same can be said for the deal on the office copy paper as timely delivery—in her mind—was factored into the transaction and an integral part of the experience with the vendor. Again, any delays in delivery could have compromised future dealings with them.

So how do businesses deal with connected consumers? It could be as simple as connecting to them in the ways they choose to be connected. In my view, the future of successful business is not necessarily about relating to consumers but in establishing strong connections. These connections can withstand challenges and threats due to miscommunication, unplanned experiences, and unfortunate situations. They're connections so strong that the consumer has a compelling reason to stay and continue to do business with the vendor. It could start with becoming a connected business.

CHAPTER FOURTEEN
CONNECTED BUSINESS

You have to build connections to your customers in order for them to buy your products or services. You can appeal to my intellect (and you should), but if you don't also engage my emotions, the sale is going to someone who does.
– Michel Hogan, brandology.com, 2008

1. Connected Small Business

Jen Harwood lives and works in a country town called Bendigo in the state of Victoria in Australia. She runs a successful speaking, coaching, and consulting business from her home office. She enjoys her remote lifestyle while maintaining connections to the local community, state, national, and international business networks. Jen knows what she wants to achieve with her business. She has specific and achievable personal and business goals that are relevant to her place in life. She's planned and compiled revenue and expense budgets that need to be met on a regular basis to keep her business sustainable. Due to her choice to reside in a remote area, it's a challenge to stay connected to her customers and prospects and her support base of family and friends. She decided to make some changes to her business and some investments in connection technologies.

Jen took up the challenge of her personal assistant's sudden resignation to transform her traditional office into a mobile business operation. She determined that adopting various technologies would assist in making her more efficient while potentially saving her money along the way. She started by transforming her telecommunications setup to work for a mobile businessperson. To handle calls when she was out of the office, she arranged for her business calls to be diverted to a professional answering service so that her customers could speak to a real person, instead of a recorded message. Callers could leave messages with the receptionist agent who would forward messages via e-mail or text message to her Palm Treo mobile device. The combination of her mobile device and her receptionist agent allowed her to be directly contactable via voice, text, or e-mail at anytime, anywhere in the world.

As she still received business contracts from customers and suppliers who used fax machines, she installed fax software that allowed her to send and receive faxes from her laptop computer. This included being notified via e-mail when faxes arrived on her Palm Treo. When she was in her home office, Jen reduced her outgoing national and international call expenses by using a popular voice over the internet application called Skype. Calls between her contacts that had Skype installed were free and calls to phones or mobiles were priced at a fraction of any carriers' offerings. From an administration perspective, she outsourced diary scheduling, order processing, and simple paperwork activities to a virtual assistant service that performed these tasks at a fraction of the cost of a casual assistant. To maximize her cash flow, she procured a mobile payment system that allowed her to process credit card payments for book purchases while she was at speaking engagements, on the spot.

The transformation took less than thirty days and reduced her costs of a full-time personal assistant from over $2,000 per month down to $200. By making smart investments in technology available to any business today and being selective with how the applications could add value to her and her business, she became more efficient at managing communication processes at reduced expense to her business. From a customer service perspective, Jen also reduced the human error of missed calls and mismanaged customer information by outsourcing noncritical process tasks, scripting responses, and thinking the customer experience all the way through. Her existing and new clients appreciated the personal service and prompt contact. By doing this, Jen had more time to work on promoting her business, developing new products and services, and spending more time with family.

She then focused on her online presence. For the customers and prospects who wanted to interact with her online, she implemented simple Web 2.0 (two- or more-way interaction) applications. She started with making it easier for people to find her and her company. Jen was already registered with the online business network LinkedIn. She updated her profile to a level that best represented the image she wanted to present to the market. She added information about her background that she thought prospects would want to know now about her prior to engaging her services. She requested and received testimonials from existing customers and participated in selected group discussion activities. By establishing an appealing profile on a global online network like LinkedIn, she increased her marketability beyond her community. She knew that she stood a better chance of appearing in Google searches of her own name and that of her business if she had a presence there. She also started to connect to customers and partners who were already on LinkedIn to develop her network of connections.

Then she looked at online marketing. Jen had already participated as a speaker in podcast interviews and conducted webinars so she was familiar with the effectiveness of connection technology to get her message across to people who wanted to use these connection technologies. She updated her web site by including links to these connection technology information sources. This complemented the capability to inquire about and secure a booking online for her speaking services. She had previously set up an order and pay component on her site to enable online purchases of her published books, *How to Get the Job U Want* and *The Art of Networking*, on a self-service basis to anyone in the world with an internet connection and a credit card, even when she was asleep.

In addition, she wanted to provide an interactive capability to get real-time feedback on the information she posted through user comments and discussion. She set up a linked blog site in less than thirty minutes one evening and began to populate it with relevant content for her target market. She gave prospects further insight into her personality by producing a short introduction video using her Apple's iMovie application about herself and her work to give interested viewers a brief glimpse of the real Jen Harwood, in her own words. She posted it on YouTube so she could send web links to people who were interested in finding out about her service prior to engaging her so that she didn't have to pay additional video hosting costs on her own web site.

Finally, she registered for a service called sendoutcards.com. This service allowed her to electronically request a customized greeting card with her own personal message to be printed and mailed to any of her contacts for less than the cost and time of buying and posting a card herself. She thought it would be ideal for sending to her relatives, friends, customers, or prospects at the appropriate times.

2. Connected Employees

Luke is a telephone sales representative in his mid-twenties and works for a large corporation. Luke is a people person who always seems to know what's going on, both at work and in his social life. He has excellent relationships with his 150 customers. He attempts to get in touch with them regularly via a disciplined contact cadence system requiring that customers be called on at least once every two months. For his key customers, he arranges face-to-face visits despite the fact that he works in a call center. For his most important clients, he proactively arranges to invite the ones he knows have an interest in sports to his company's corporate box at the local sports stadium. His customers appreciate that they can call or send him a text message on his mobile phone at any time of the day or night to inquire about any aspect of the company's product and service offerings.

Luke discovered early on that his mobile phone is a more effective sales enabler than his desk phone. He uses it to stay in touch with his customers—as well as his friends—and leaves it on twenty-four hours a day to support his commitment that he will take calls from his customers anytime. Should he be in a meeting, traveling, or otherwise not contactable, his preference is for callers to leave a text message so he can respond instantly via a return text, although he generally responds to voice messages within thirty minutes of the message being left. They show their appreciation of Luke's relationship with them by scoring him highly in company customer satisfaction surveys regularly, where he consistently tops the list of account managers with the most loyal customers.

It comes as no surprise that Luke overachieved his sales targets for the last three years. For business, you can usually find him on the online network LinkedIn, where he manages his relationships with his customers, business partners, suppliers, and colleagues. Socially, he is usually involved in either organizing or attending social events for work, which includes team-building sessions and drinks down at the local bar. He is also a regular on Facebook, which he uses predominantly for social activities but also for work when appropriate.

Luke's employers, while grateful for the efforts he puts in based on the customer satisfaction targets and his sales budget attainment, are not familiar with this style of conducting business. As a call center worker, Luke is expected to be at work and start making outbound phone calls from 9:00 a.m. until 5:00 p.m. with a break for lunch. This is the tried and tested corporate telesales model and they believe that there should be no variations to something that's worked well in the past. Luke requested more flexible working arrangements to suit his customers and his lifestyle. He asked to work from home one day a week to care for a sick family member. In his request, Luke detailed the benefits of a more mobile arrangement, which included a minimum three hour per week productivity benefit to the company based on travel time saved alone. While he was granted approval to do so on a trial basis, his management is wary of the precedent it sets for other workers at the center.

3. Connected Big Business

American retailer Wal-Mart has been struggling with its brand image in recent times with many people questioning the company's labor practices, purchases of Chinese products, and the impact these decisions have on Americans. These activists, who gathered on sites like walmartwatch.com, were online savvy and ready to pounce on any information, truth or rumor, and share them widely with their community network and the media to validate their cause. Wal-Mart needed to address the reality of having to deal with these detractors without positioning itself as the powerful corporate ogre while continuing to serve the hundreds of millions of satisfied customers who regularly shop at their stores.

The company made a conscious decision to move into the social media space to provide a right of response, an ability to put the company's position forward and to listen and learn from customers who choose to use this medium. They began a strategy of informally participating—with no corporate communications spin—in various consumer blogs in a voice that resonated with customers. They figured that with more Americans becoming familiar with connection technologies, Wal-Mart could use supporters to counterbalance what detractors were propagating. Lisa Everett from BNET's *Groundswell* quotes Lucretia Pruitt, a panelist on the elevenmoms.com consumer blog, who said, *"I've dealt with many*

C-level execs in my lifetime (as well as many attorneys) and I've never seen such openness to an idea that is clearly going to rock the foundations of 'how business has always been done.'"

4. SCAN test

In previous chapters, I've discussed the concept that connected individuals innately exhibit *HITS* behavior—*hunt, interact, test/trial, and share*—in an online environment. As members of a connected group, they instinctively *GROUP*—*gather, regulate, organize, unite, and participate*—in the areas of interest that directly impact them enough to function collectively. As businesses can be an individual—as in a self-employed arrangement or a group—or a small, medium, or corporate business, their connection behavior mirrors their individual or group connection roles. However, they also possess another set of factors that defines them as connected businesses, which is revealed via a SCAN—*smart, customer focus, accessible, and nimble*—test.

SMART. This concept has been covered in previous chapters. It's about having *specific, measurable, achievable, relevant, and timely* goals for a connected business. As an individual small business operator, Jen worked on her specific personal and business goals regularly to achieve the outcomes she desired. She realized that this was about both work and life, which is an important distinction that many fail to make. She knew the key to achieving her goals was to set up milestones that would allow her to measure her progress.

As for Luke, he already had his metrics for success set up for him by his employers—sales budget for the year and regular customer satisfaction surveys. There was always an annual debate between management and the sales representatives about whether the budgets were achievable and realistic or not, but at the end of the day, Luke had to find a way to make his targets. Also, Jen had to work on the challenge of finding alternatives to a personal assistant that were still relevant to her business and frame her goals with a time element. For both Jen and Luke's businesses, it was an annual sales measurement, with regular milestones set on a monthly basis. As simple as this thinking might sound, a successful connected business can move to the next part of SCAN once it has invested time to plan for the *SMART* part of the equation.

Customer Focus. For the connected consumer, trust and relationships are important. To become a successful connected business, the adage that the *"Customer is King"* has never been truer. Wal-Mart realized it and decided to participate and listen to customers before the customer perception of the brand was influenced by detractors. Jen was convinced that her customers wanted to speak to a real person so she made arrangements through service providers to ensure they received this experience. She also invested in a service that allowed her to send out personal messages using traditional greeting cards, as she felt that customers would appreciate a personal communication touch. At the same time, she also set herself up for customers who wanted to interact with her online by establishing business profiles and video messages on online sites, podcasts, and webinars for those who preferred to receive information via this medium with blogs for information sharing and feedback. The main focus was about how they chose to connect to her, not how she chose to connect to them.

In Luke's case, he knew that providing a service to contact him at any time about any thing would cause him more headaches as he fielded calls about billing, service, and delivery issues as well as a fair share of complaints. He also knew that each of these calls was an opportunity. An opportunity to better relate to a customer by addressing problems or by acting as the point person within the company to prevent the customer from getting frustrated with the potential bureaucracy of dealing with multiple departments to get their issue or complaint resolved. Luke understood that he was more than a customer representative by job title. It was his accountability.

Accessible. In a connected world where information is a mouse click away, connected businesses need to be where the customer wants them to be. In many cases, this means accessible when he or she requires them. This doesn't mean contactable necessarily—this means to be present in a form that's acceptable by the customer. There's the example of the remote-controlled garage door that decides to malfunction by opening and closing intermittently in the early hours of the morning, which, despite all efforts to correct, the owner cannot repair. As this isn't an emergency, it would be extremely difficult to hunt down a twenty-four-hour garage door repair person or even track down an all-hours phone number to call for such incidents. However, a company web site that has a frequently asked questions (FAQ) capability that provides tips and advice on how to disarm automatic garage door devices, would probably fix the problem.

The point is about being there when the customer wants to connect to you. Jen established presences on sites like LinkedIn and YouTube so she could be found more easily on Google searches of her name or that of her business. She also ensured she was able to receive information and messages relevant to her business via voice, text, e-mail, and fax at any place, any time—on her mobile device. Luke didn't need convincing of the value of being available on his mobile phone at any time. The satisfaction ratings his customers regularly provided and the sales results he consistently achieved were proof enough that accessibility was a critical requirement for his customers. Neither did Wal-Mart; if the company wasn't actively seen online, it wouldn't be heard either.

Nimble. I've worked for two large corporate organizations over twenty-two years. When I'm asked what the one constant factor I experienced during all that time is, my answer is: *change.* I worked at IBM in the early 1990s when the company was on the verge of irrelevancy with its stocks plummeting and customers leaving in droves. I was at Australian telecommunications giant Telstra in the mid-2000s when the company had to transform itself from the shackles of government ownership to become a viable and credible privatized entity that shareholders wanted to invest in. To be a connected business, you need to be flexible and nimble enough to change with the times and the reality of business environments—or be prepared to potentially perish. The good news is that connection technologies are part of an ever-changing business environment due to their innovative design and nature. These technologies are developed based on improving process and communication and are the enablers required for change.

Jen knew that staying in a static office may have been relevant to her previous business goals. However, they were no longer relevant to someone who wanted to expand her business beyond a remote community and reach out to domestic and international markets. She needed to be nimble enough to transform her business to address the business need of expansion while organizing for the practicality of being a mobile worker.

Recent studies show that businesses are factoring in mobility. A 2008 International Data Corporation (IDC) Asia report found that 81 percent of managers believe working from home arrangements improve productivity, which is attributed to the variety of tools that connect mobile and remote workers such as video-conferencing and web-based office applications like Google Docs. Luke hopes that his company can change with the times not only to provide better employee flexibility, but so they can realize the productivity benefits at the same time.

5. Business Connectedness

There is much discussion amongst business analysts as to the true value of what they refer to as social media, new media, Web 2.0, and other variations of what I call connection technologies. The references are usually about how these technologies are about "communication and conversation." If this is the case, they've all missed the point. Business models are evolving from a "process" of constant communication to the quality, degree, and types of "relational interaction" between vendors and supplier. More human factors will come into play relative to the "relationship" mix as we move to greater business "connectedness."

Businesses MUST find a way to CONNECT with their prospects and customers through a mutually acceptable two-way interaction platform. The future successful business will be more about the business finding a way to relate to the prospect and customer via a tangible social, business, and/or human connection. It can be as simple as Googling a prospect's name and identifying the exact individual you want to target. Once the correct profile is located—usually through an online presence or blog—it should reveal some basic information such as who they are, what's their expertise, what their business interests are, and what they like to talk about. If they are on an online network, one can hunt down more specific facts about their business network and potentially some social information. Marketing cynics may argue that this isn't feasible in big markets due to the volume of customers. The point is that if every business knew its top one hundred customers through simple information that is easily searchable online, the business stands a better chance of retaining them longer. Businesses that have a tight social-business-human relationship with their customers will be best positioned to acquire, win back, and retain those customers over a longer period of time.

The other dimension to this is a connected business metric. The common method of measuring the number of retained customers is not sustainable in a connected world. As customers gain greater access to product and service information, partners, suppliers, and ideas through the web, businesses need to become "connection-centric." The question that needs to be answered is not how many customers businesses have but *HOW CONNECTED* are they to their customers. Connected as in: how well you know them—they know you, they trust you—and how satisfied they are with your business relationship and loyal to your business.

In an increasingly connected world, the measure of success will be about how many customer "connections" you have, and the strength of this connection will determine how many products and services you can sell them—and continue to sell them—over the period of the business relationship. On the other hand, if you have tenuous connections with your customers, they are likely to switch allegiances to a vendor who is willing to invest in a business relationship that connects their business needs to their vendor of choice.

CHAPTER FIFTEEN
CONNECTED LEADER

The point about Connectors is that by having a foot in so many different worlds, they have the effect of bringing them all together.
– Malcolm Gladwell, The Tipping Point

1. Connection Campaigning

As an interested observer residing in Sydney, Australia, I was intrigued by the choice Americans had to make in November 2008 as to who would be their next president—and the leader of the most powerful nation on the planet. An interesting consideration was the electoral system in the United States where the choice is not only between two (sometimes more) presidential candidates or preferred political parties. Due to voting being non-compulsory in the United States, the choice adults had was more about whether they could be bothered to cast a vote or not.

As a marketer, I was fascinated by the communication campaign that was unfolding in the months leading up to the election. It was analogous to a marketing campaign developed to influence consumers to become aware and interested about a branded product or service and its benefits, and, more importantly, create the desire and preference that would lead to the physical action of "buying" it. Traditional communication such as newspapers, magazines, television, and the face-to-face public gatherings and town hall meetings were always going to feature as they had in previous campaigns. What was enthralling was how presidential candidates were going to leverage online to manage their campaigns beyond communication.

In standing on the virtual sidelines thousands of miles away from the election campaign, I could relate to how people who had distanced themselves from the political process previously could now become more exposed to what the candidates and their parties had to offer, simply by accessing the various connection technologies at their disposal. How the candidates would go beyond communication and, in fact, establish a connection to their constituents that would lead them to not just read and listen to the rhetoric but to vote with their feet (walk to the polling booths) and fingers (make their choice on a computer screen using e-voting) on the day.

In the lead-up to the election, I watched how Barack Obama, the *super* connector, and John McCain, the *basic* connector, ran their respective campaigns in a connected world. The winner would not only become the next president, he would also become a *connected leader*. The three basic factors that determine the degree to which a connected leader's attitude and behavior toward connecting to people and information are the *Three Rs of Connection—Reason, Relationship, and Return.* An assessment of each of these factors unravels some clues to how McCain and Obama approached their online campaigns relative to connecting with their constituents.

2. The Three Rs

John McCain experienced the power of the web in previous campaigns. He knew he could use the web to raise funds and rally support for a cause. In 2000, his online campaign had successfully raised more than six million dollars in campaign funding; in 2004 the Republican National Committee used the web to assist with organizing around 1.4 million volunteers to participate in a get-out-the-vote operation. His team had tried and tested the technology and it seemed that it was only a matter of executing a new strategy that leveraged the web.

In Barack Obama's case, he did his research. According to the Pew Internet & American Life Project survey conducted in April 2008, almost 50 percent of Americans used the web to follow the presidential campaign and around one in six (17 percent) browsed political web sites, read campaign e-mails and text messages, and generally used the internet to stay updated with the election on a daily basis. Sixty-five percent of Obama supporters said they followed politics online, compared to 56 percent of McCain supporters. The survey found that 35 percent of Americans watched political videos online, 10 percent used social networks to become involved in the campaign, and Obama's three target constituencies—young, black, and affluent voters—showed the highest growth rates in online news consumption.

Obama studied Howard Dean's 2004 campaign as well as George W. Bush's last two successful efforts in developing his own 2008 campaign strategy. An article in *Time* magazine in June 2008 mentioned that one of Obama's three ruling campaign principles was to "build it from the bottom up." *Time* magazine's Karen Tumulty writes: "*Obama's Chicago Headquarters made technology its running mate from the start. That wasn't just for fund-raising: in state after state, the campaign turned over its voter lists...to volunteers, who used their own laptops and the unlimited night and weekend minutes of their cell phone plans to contact every name and populate a political organization from the ground up.*"

Obama wanted to build a sustainable network of supporters who organized themselves and marshaled others to action. In the end, McCain's camp decided to use the web based on historical precedent, which was primarily to raise campaign funding and to communicate to supporters. A month before the election, *Time* magazine writers summarized Obama's online plan as "*...seamlessly integrating the networking power of technology with the flesh-and-blood passion of a social movement... (by using) ... the greatest tool ever invented for connecting people to others who share their interest.*"

From a relationship perspective, both candidates understood the importance of establishing rapport with their constituents. They also knew that in May 2008, research firm Parks Associates reported that 82 percent of Americans were now online, compared to the 41.5 percent in 2000. McCain's political advisers hired four consulting firms to develop an interactive online campaign plan. They realized that McCain was playing to an older base of supporters but still developed an online presence dedicated to his web-savvy followers. They built a basic web site, featuring blogs, a social networking feature called McCainSpace, and even a game application named Pork Invaders—styled after the 1970s' Space Invaders arcade game—that let you fire bullets marked "*veto*" at slow-moving pigs. His staff also posted a video advertisement on YouTube called "A Man in the Arena" that celebrated his wartime service. His daughter Meghan and two of her friends wrote a "bloggette" from the campaign trail.

On the other side, Obama was willing to use every connection technology available to form sustainable connections with and among his supporters. His camp custom built interactive web sites such as mybarackobama.com that included click-to-donate tools and its own built-in social network registry. He recruited Chris Hughes, cofounder of Facebook, to help develop and manage the online campaign. Obama established comprehensive profiles on all the major social networking platforms including MySpace, Facebook, LinkedIn, YouTube, and Twitter and actively encouraged e-mailing, blogging, text messaging, posting videos, audio recordings, podcasts, and even direct feedback.

In the early days of his campaign, his team set up a customer service center staffed by real people who took calls twenty-four hours a day. Obama's supporters created an unofficial think-tank web site called Oh Boy Obama! for the purpose of giving their "grassroots" supporters a platform to submit and vote on ideas—as they happened—to provide guidance to Obama's campaign team on key topics and issues relating to the election campaign. Finally, there was no doubt as to what the return would be for their online efforts. In this case, the prize at stake for both candidates was obviously the presidency of the United States—and the mandate of the people.

In the true form of a *basic*, McCain delegated his online campaign to people he assumed could act in the best interests of what he was trying to achieve. Being a *super*, Obama managed his campaign in a disciplined fashion, maximizing the power of the individuals and groups within his ever-expanding network In April 2008, *Time* magazine's Michael Scherer and Jay Newton-Small quoted Republican internet consultant Mindy Finn on Obama's online advantage: *"Everything Obama does is fundamentally about a people-powered democracy and a people-powered campaign. McCain's message is different."*

3. The Results

McCain's online plan never got off the ground. By October 2007, most of McCain's online employees had been fired. In January, McCain admitted on politico.com that he relied on his wife for assistance with computers due to his "illiteracy" with the technology. In April 2008, the official McCain web site was described by *Time* magazine as *"the technological equivalent of a soapbox-derby car on a busy freeway."* It was noted that the campaign blog was not updated regularly and McCainSpace was not finished, with supporters being directed to "stay tuned." The site had a page that asked for money, detailed policies, and posted issues papers, press releases, and videos. A former adviser to the McCain campaign was quoted as saying, "*…you knew at a certain point that it wasn't going to be the kind of online campaign that had been planned.*"

In July 2008, McCain confessed to the *New York Times* that he didn't use e-mail and that people accessed the web on his behalf. In September 2008, McCain's camp was further embarrassed on the online front when vice presidential running mate Sarah Palin's private e-mail account was hacked.

The hacker then revealed the relatively simple steps he took to crack the e-mail account by simply resetting Palin's password using her birth date, zip code, and the security question where she met her spouse, which was found with a simple Google search. The campaign resembled an unplanned and unstructured online campaign.

Obama's campaign was about the people and engineered for engagement. He ran one of the world's most astute interactive campaigns in history. The results speak for themselves:

- According to internet traffic tracker Hitwise, he regularly attracted around 68 percent of all weekly presidential campaign-related web traffic.
- According to data marketing company Compete, he regularly generated 2.6 million weekly unique visitors to his official web site.
- A few days before the election, he had amassed 740,000 MySpace friends, over two million Facebook fans, and 105,000 followers on the online networking platform Twitter.
- A hip-hop music video set to an Obama "Yes We Can" speech was posted repeatedly on YouTube with the top two postings alone being viewed ten million times.
- By November 4 (election day) he had accumulated 126,930 subscribers on his YouTube channel with 19,114,541 views of his video content, compared with 29,318 subscribers and 2,137,815 views for his rival McCain.
- One of the most popular Obama YouTube videos is a clip of him tripping over a roadside curb while reading a message on his BlackBerry device.
- In the last few weeks leading up to the campaign, the Obama camp launched an iPhone software application

called Obama '08, which provided users with *"instant access to Barack's positions on important issues, as well as local and national campaign news as it happens"* and he bought in-game billboard advertising space on nine Xbox360 games to directly target the eighteen- to thirty-four-year-old male demographic.

- Throughout his campaign, he raised around $600 million in campaign funds over the web, mostly in microdonations from masses of individual supporters.
- He won 68 percent of the votes of those who voted for the first time, mostly young people.

McCain seemed to be stuck in the old world of campaigning. He attended the rounds of hotel banquets where he charmed wealthy supporters with lavish dinners while Obama was winning the connected voter almost one at a time. Scherer and Newton-Small from *Time* magazine noted that in March 2008 alone, *"(McCain) attended 26 fund raisers in 24 cities, raising about $15 million with roughly one-third of it coming from the web. Obama attended just six events in the same period, yet his campaign raised three times as much, mostly online."*

McCain should have paid a little more attention to the research on online usage profiles. The McCain camp claimed that his constituents didn't spend much time on web applications and online tools because their base wasn't comfortable with the technology. While the Pew Internet & American Life Project research found that only 35 percent of Americans over sixty-five are online, the subset that matched McCain's race, gender, and education was 75 percent.

Obama knew he had to target the youth, black, and affluent voter to give himself a good chance in the polls. On the campaign trail, he let the cameras roll while he played basketball with local college kids, hit the phones with volunteers, and danced with gay icon Ellen Degeneres on her television show. And these weren't just stories—the video records of these activities were posted on YouTube for all to see and share. He sought and gained the public support of the most influential black Americans in Oprah Winfrey and political leader Colin Powell. By November 2008, he was ranked the fifteenth-most connected individual on business online network LinkedIn, which numbered thirty million users at the time and consisted of predominantly senior North American business professionals and executives who earned an average $105,000 per year.

4. The *OPEN PLAN* to Make History

It's now history that on November 4, 2008, Barack Obama was elected the forty-fourth president of the United States of America with many analysts agreeing that his online initiatives were an important part of his victory. Julie Germany, director of George Washington University's Institute for Politics Democracy & the Internet observed that *"No one's going to say Obama won the election because of the internet but he wouldn't have been able to win without it. From the very beginning the Obama campaign used the internet as a tool to organize all of its efforts online and offline. It was like the central nervous system of the campaign. A lot of what (the) campaign did online was trying to encourage people to do something for the campaign. It was not about passively absorbing information."*

Micah Sifry, cofounder of techpresident.com observed that *"Obama understood the power of the network that he built to support his campaign. He understood the power of individuals self-organizing in support of his campaign."* Obama ran his online campaign like a business. He treated his supporters as customers and advocates for his brand. He executed his strategy with military precision and, in my view, followed the *PLAN* to manage connected individuals for the ultimate payoff. He *personalized* his appeals by actively connecting to people. His multitude of MySpace and Facebook friends volunteered by opting-into his fan pages. His LinkedIn profile is written in a personal style as if he wrote it. In contrast, McCain's is written in the third person.

Obama utilized a range of connection technologies that provided the "*listening* devices" to what his supporters were telling him, including campaigning by phone, canvassing, e-mails, blogs, instant messaging, social networks, videos, podcasts, text messaging, twittering, and a range of interactive web sites. The people chose how to provide feedback to his campaign initiatives and he made it easy for them to find him. He *activated* his connections to action. While both candidates built online tools designed to generate easy-to-use lists for volunteers to make calls or knock on doors, more people talked about and used Obama's.

Micah Sifry noted that Obama's Neighbor-to-Neighbor initiative returned 479,000 hits when searched on Google compared to a mere 325 hits for McCain's Voter-to-Voter equivalent. The record voter turnout is also an indicator of this "activation" of his followers. Finally, throughout the campaign, Obama *nurtured* his connection relationships. He made fund-raising a social event, giving supporters personal targets, letting them run their own events and watch fund-attainment thermometers rise. As a registered fan on his Facebook page, I received regular invitations and updates on events, debates, announcements, and appeals—all from my laptop in Australia.

From a group perspective, he had an OPEN outlook. Although untested, he knew what the *opportunity* could deliver by marshaling his supporter base for the cause, although he underestimated it. Obama admitted in June 2008 that he *"...didn't anticipate how effectively we could use the internet to harness that grass-roots base, both on the financial side and the organizing side."* He got people actively involved and *participating* for the cause.

There's the story of the couple from Ephrata, Pennsylvania, who, in January 2008, accessed Obama's web site and created a personal web page to connect to other Obama supporters in their area. The virtually organized group began with an offline meeting were they assigned each other tasks based on guidance from the web site, including downloading phone lists to cold call potential supporters and typing names and addresses into national campaign databases. When Obama's paid operatives arrived to conduct training sessions three months later, two thousand people were ready for them.

Obama's campaign was based on *engagement.* Jascha Franklin-Hodge, the chief technology officer of Blue State Digital, the company that managed Obama's digital resources said, *"Millions of phone calls were made to early primary states by people who used the web site to reach out and connect with them. We facilitate actions of every sort: sending e-mails out to millions and millions of people, organizing tens of thousands of events...the point of the campaign is to get someone to donate* money, make calls, write letters, organize a house party. The core of the software is having those links to taking action—to doing something."

He successfully tapped into mobilizing the Obamaphiles, especially on the issues that could impact the campaign.

In the year leading up to the presidential election, Obama's campaign was subjected to rumors about his faith, patriotism, and even wife Michelle's opinions on race. To assist with identifying and addressing these potentially damaging rumors, Obama launched the fightthesmears.org web site, which offered official responses to rumors and encouraged supporters to post responses and notify the campaign team via e-mail if they heard of any "new" rumors. The site kept a close virtual eye on baseless reports that could demotivate his supporter base by asking them to keep their virtual ears to the ground for him.

Finally, there's good old *networking*. Obama's camp was proficient at using social and business networks as active agents of his campaign to not just act—but incite others to act too, which I call sharing with a purpose. According to *The Diplomat* magazine's Nick Bryant in Sept. 2008, *"(Obama's) chances of becoming commander-in-chief have been enhanced greatly by being America's social networker-in-chief."*

5. SMART Interconnectedness

In a *New York Times* article on November 15, 2007, Roger Cohen said, *"Obama, in many ways, is where the world is going. He embodies interconnectedness..."* Obama did his homework, designed, and delivered the most efficient and effective connection campaign that achieved his stated goal of victory through the people. He used connection tools and leveraged the talent of his supporters to realize his purpose through interaction and participation. He ran a *SMART* campaign. He had a *specific* goal and surrounded himself with the people, groups, and tools that had the interest and expertise required. He *measured* his progress at every step in managing his connections, fund-raising targets, and, ultimately, votes. His online plan was *achievable* as he set realizable goals and outcomes for the campaign. His theme of "change" was more than *relevant* to his constituents at the time and his ability to connect individuals, groups, and networks was *timely* at every step of the campaign all the way to the moment of his victory.

The last word belongs to Obama the connected leader in an extract from his victory speech on the evening of November 4, 2008:

"I was never the likeliest candidate for this office. We didn't start with much money or many endorsements. Our campaign was not hatched in the halls of Washington—it began in the backyards of Des Moines and the living rooms of Concord and the front porches of Charleston. It was built by working men and women who dug into what little savings they had to give – five dollars and ten dollars and twenty dollars – to this cause. It drew strength from the young people who rejected the myth of their generation's apathy; who left their homes and their families for jobs that offered little pay and less sleep; from the not-so-young people who braved the bitter cold and scorching heat to knock on the doors of perfect strangers; from the millions of Americans who volunteered, and organized, and proved that more than two centuries later, a government of the people, by the people and for the people has not perished from this Earth. This is your victory."

CHAPTER SIXTEEN
LEARNING CONNECTION

*It could become a World Wild Web. Let us all make it a World
Wise Web.*
*– Jean-François Abramatic, W3C Summary Presentation, April
1997*

1. Same Old, Same Old

The last academic lecture I attended was in 1982. The subject was mass communications and I needed to pass it to
complete my bachelor's degree. It was not a particularly
memorable event—the lecturer arrived at the venue to a
gathering of around forty students, proceeded to the lectern, powered up the overhead projector, and began speaking to what seemed like a multitude of prepared overhead
slides in a monotone voice, ironically about the future of
mass communication technology and something called the
internet. We listened passively and took notes. It must be
said that if grades were given out for the lecturer's success

in effectively transferring his knowledge on the subject to the student audience, he would have received low results. In fact, I would rate him a "Fail."

Fast forward a full twenty-six years to 2008 and an opportunity to attend a university lecture. I accompanied my son to the first week's lecture of an information technology subject he was undertaking in the new semester. I envisaged that with the step-change developments in presentation training and technology over those years, I would have an infinitely more enlightening experience than the last time around. I watched as the lecturer entered the theater-style hall to a mass of students numbering around six hundred, approached what looked like a flight deck of computers and light switches, pinned on his lapel microphone and checked that it was functioning, and reminded the audience that the lecture was being recorded in podcast format and, with the soft-copy version of the presentation slides, would be posted on the course web site and available for download within two hours of the presentation. He then accessed his standard set of PowerPoint slides and presented what seemed like a hundred slides in an eerily familiar monotone voice that had students slowly exiting the hall within ten minutes. I also observed some students so bored that they preferred to send text messages on their mobile phones and even saw one student playing a game on his PlayStation Portable device. I valiantly stayed for the scheduled ninety minutes taking written notes as I'd done all those years ago. I left the building with around forty students, which was a coincidental reminder that nothing much had changed in the lecture delivery. Nine weeks into the course, my son went to the scheduled weekly lecture but it was cancelled five minutes into the session as only seven people turned up.

My understanding is that this university experience of face-to-face education delivery would be generally similar to current practices in primary and high school classrooms and even corporate organizational training. Although there have been significant initiatives by educational organizations to provide more learning options using connection technologies, it seems that the simple transfer of material from a traditional medium to an electronic one and presented in the same style is a common practice and relatively ineffective in knowledge exchange in a connected world. This observation begs three questions: Are connected learners different from students of the past? If so, how do educators plan and deliver more effective learning? What are the recommended methodologies to effect this?

2. Learning HITS

Connected learners comprise a subgroup of connected individuals. As this book consistently maintains, the attitude, behavior, and capability of individuals in the *Connection Generation* are fundamentally different from those who preceded them. This is exemplified by looking at how the connected individual behavior pattern of *HITS—hunt, interact, test/trial, and share*—fits with the connected learner role.

Hunt. Successful teachers and students have always been good resource hunters as it's their job. Whether it's researching topics to deliver as part of a curriculum or preparing for a course assignment, research is critical. With the advent of search technology, this activity was facilitated by the range of topics available to be searched on the web, enhanced by the depth of information that was posted on web sites and compressed by the speed with which research work could be accessed and tracked down. Information sources that are available online go beyond the familiar written form and are more easily accessible. In the past, if an academic wanted to track down the historical audio or video record of John F. Kennedy's assassination in Dallas in 1963, he or she would have to go through a protracted process of obtaining the footage. Today, anyone can see this same footage digitally on YouTube and source additional pictures, audio, and video of witness accounts of the event contained only in independent documentaries through a range of web sites. Some sites even did the research work for academics. Although sites such as Wikipedia are seen as an unreliable collection of factual data, students regularly use the references section as a valuable index of pointers to more accurate sources. In addition, online networks give these hunters the option of directly hunting down the subject matter experts once they are identified by a simple name search so they can be contacted and connected to directly.

Interact. Once they hunt down the information, they want to interact with it. Not just read, listen, and watch but cut-and-paste it, download, and present-and-publish their own perspectives of what the information means to them. They want the information to provide meaning to them in their way—not in a way that is dictated to them. Events such as the 9/11 terrorist attacks in the United States in 2001 and the Bali bombings in 2002 were not events that happened in far-off lands that were interpreted by Western media on traditional mediums such as television, newspapers, and magazines. People want to understand these events from sources who were closer to the action or who had other points of view that were not voiced through mainstream broadcasters. They now have access to networks such as Al-Jazeera for a different perspective of the events that they would not have necessarily been directly exposed to. They also want to interact with each other to make sense of the information. They can go beyond the thoughts and ideas of their own circles of influence and read, listen, and watch the myriad of global opinions contained in blog posts and forums from all over the planet.

Test / Trial. Connected learners are more willing to try out new connection technologies that will assist them in getting better connected to people and information. The preference of downloading ilectures, recordings of previously delivered lectures available on course web sites, is increasingly common. So much so that by June 2008, the Apple iTunes University education site, describing itself as the campus that never sleeps, provided free video and audio content from three hundred worldwide educational institutions. It plays to the user preference for flexibility, portability, and range of information as well as the capability to find lecture content quickly in one place, without having to wade through a single university web site.

Teachers are also willing to be trained to support connection technology. In October 2008, a survey by Australia's national agency for the integration of technology in teaching, education.au, reported that 79 percent of teachers responding to the survey said the internet was an essential part of their job but only 36 percent claimed to be proficient and confident. The survey also showed that teachers embraced technology as a factor that would "enhance their teaching and professional development."

Share. As part of her university geology major, *active* Rachel had to complete some field work in the form of two weeks in the remote town of Plumbago, South Australia. The field trip was coordinated through the collaboration of Sydney and Macquarie universities, bringing together sixty students completing geology studies. Due to the remoteness of the location, the nineteen- to twenty-five-year-old students had to deal with no internet, phone, or mobile coverage for fourteen days—in fact, no communication with the outside world. To this age group, this can be viewed as tantamount to some kind of torture and evil depravation. Although they began as relative strangers, they all got to know each other and gradually formed relationships.

When it came time to leave the group and head back home, there came the sudden realization that they weren't going to see each other very often—or at all—post excursion. The group made a collective decision to keep in touch by committing to interaction via membership on a Facebook group page. On the group page, they could continue conversations and inside jokes through wall postings, share photos and musings from the trip, and maintain direct engagement with all group members. The Sydney and Macquarie University Geological Society, or "SMUGS," was formed on Facebook upon their return and almost immediately, students, lecturers, tutors, and support staff joined the group. Within a week, over three hundred photos were posted from various members and countless comments posted on the group wall. The group was planning to organize a "big night out," which is now facilitated by an online network and a committed group of students who ironically started out as strangers and had no online connectivity capability.

Along with their hunting, interaction, and testing behavior—and despite the competitive nature of the education process—students are willing to share information. Apart from blogs and forums, online discussion groups have become prevalent due to their ability to discuss and share information, thoughts, and ideas in a less formal and unregulated environment than a classroom or tutorial learning format. In addition, what's shared may be more than what's being learned today. The SMUGS fully realize that the connections they made on a field trip may be as valuable to them today as it could be years from now as commercial geologists.

3. Learning PLAN

As connected learning behavior is equivalent to that of connected individuals, it makes sense that educators should plan and deliver more effective learning through a methodology that works for connected individual behavior. This comprises the four factors discussed in previous chapters of *personalize, listen, activate, and nurture*, and best summarized as a *PLAN*.

Personalize communication. Connected individuals, whether they are students, teachers, lecturers, or trainers all have a connection profile. *Basics* and *supers* are as different in terms of how they relate to people and information as chalk and cheese, as are *passives* and *actives*. Even *selectives* have their own way of managing connection activity. It's imperative that educators have a good understanding of each of these profiles to best maximize learning outcomes and, as much as possible, design and deliver course material in the way each profile chooses to learn. This understanding could lead to identifying connection technologies that are more effective in knowledge transfer and retention for some profiles as opposed to others.

It could also mean appealing to a sense that's more universal, like emotion, as was the case with Randy Pausch. Randy was a computer science professor at Carnegie Mellon

University in the United States and in September 2006, at the age of forty-seven, he was diagnosed with pancreatic cancer and given six months to live. He was asked to deliver a lecture on how to summarize what was important if you were about to die. His lecture was called "Last Lecture" in direct reference to the subject but also to the fact that, due to the extent of his illness, this could have been the last time he addressed the student body. The lecture was delivered using conventional PowerPoint technology at the university lecture hall. As an afterthought, the lecture was videotaped and posted on the web for people who could not attend on that day. Due to the nature of his situation and the emotion generated from the fact that the question was not hypothetical for him, he chose to relate his knowledge, thoughts, and feelings despite his impending fate and the realization that he was leaving behind a young family. The content, as morbid as it may have been, was delivered in a memorable way and shared with the whole world via YouTube, reaching ten million views by July 2008. Randy passed away that July but left a legacy of a personal connection to a global audience of millions for a lecture posted on YouTube and a book that became a worldwide bestseller.

Listen. It's imperative to listen to what connected learning individuals are saying. They may comment in a discussion forum format or post a diatribe on a blog. They may even vote with their feet as the information systems lecture attendees did by avoiding the lecture hall en masse despite being somewhat interested in the first few weeks. Did it really need to take nine weeks before the lecturer realized that his students weren't finding his lectures of value to the extent of not turning up? If he'd been watching their behaviors, asking the right questions, and listening to the feedback, he would probably have modified his delivery technique if he wanted to keep his audience.

Connected individuals are prepared to comment and vote on web sites such as ratemyprofessors.com. This web site is the largest online listing of collegiate professor ratings, with more than 6.8 million student-generated ratings of over one million past and present professors on attributes such as helpfulness and clarity in over six thousand schools across the United States, Canada, England, Scotland, and Wales. It's revolutionized the way students provide feedback to universities on educators—and more importantly, how educators are using the data to recruit, performance manage, and redeploy its teaching staff. It sets a precedent for both active listening and engaged learning.

Activate their senses. Developing an understanding of the mediums learners prefer to use is important to activate the appeal of absorbing information. A simple observation at the information systems lecture would have shown that a significant part of the audience wanted to interact by mobile device text messaging (and even a Playstation Portable). It would be an option to better engage the audience by asking them to send text messages during the lecture in real time on what they wanted to hear more (or less) about and whether they found the information interesting or boring to a facilitator who could prompt the lecturer to adjust his delivery to better suit audience feedback. This is an example of activating the audience through direct engagement with a connection technology they prefer to use. In Randy Paunch's case, you didn't even have to be physically present at the lecture to appreciate his content and messages. He used the same tools that are available to any lecturer in the Western world. The difference was that he activated the senses of the audience with his passionate delivery. No matter how many times you view it, it's obvious that he engaged the audience and the people felt a relevant connection to what he was saying and suggesting. The world was lucky enough to have had the opportunity to view the event in a virtual but nonetheless engaging and memorable environment. In this case, the medium wasn't the message. The connection was the message.

Nurture. Learning doesn't stop after one lecture. In doesn't stop after finishing a course or degree. It doesn't even stop after leaving an educational institution. Learning never stops. The human being is designed to seek knowledge and information whether gathering or hunting for it. As we are dealing with hunters who have a preference to interact and trial new things, it's imperative that learning experiences are nurtured to expand to further development. These experiences should be designed as part of a work or life development strategy that incorporates past, present, and future learning. This is a critical success factor in maintaining engagement and sustainably managing learning outcomes for the connected individual.

4. Inline Learning

To better deliver knowledge to connected individuals than a mere migration of traditional material to a more modern technology, the *Inline* factor discussed previously comes into play. U.S. research in 2008 showed that students didn't want technology to replace classroom teaching but to be complementary and in concert with face-to-face teaching. Learning that's designed for relevance, delivered to engage, and adapted to fit real-world scenarios needs to factor in the most effective connection methodology that is preferred by the parties concerned relative to what needs to be achieved, which is described as the *inline* factor.

Learning fit. Learning is most effective when it's relevant to the audience. In the case of the information systems lecture example, the six hundred-odd attendees of the first lecture realized that if the lecture slides and audio were available to them to listen to within two hours of the lecture being available, they questioned the requirement of physically attending. The option presented to them was to travel to the university and aim to arrive at the time set for the lecture, find a seat in a crowded hall, listen to the lecture, and travel back home ninety minutes later. Or they could download the lecture audio and presentation files two hours later and listen to the material whenever, wherever they wanted. This is a key reason why by the ninth week of the term, only the faithful decided to make an appearance.

Lectures that continue to be delivered in this format are doomed to fail in this *Connection Generation*. Lectures from all over the world are being packaged and distributed on platforms that users can relate to such as the case with iTunes. Australian airline Qantas partnering with global business consultancy Deloitte provides passengers on any travel class on some of their flights the choice of accessing selected video lectures of five to forty minutes in length from United States universities Harvard and Stanford and Australia's Melbourne Business School. The system was integrated with their fleet of A380 airplanes in November 2008 and the partnership is looking to run the content of BlackBerry devices in future.

Delivery. Connection technology is not the silver bullet to engage the connected individual. In October 2008, Australian educator Simon McIntyre from the University of New South Wales observed that pure online learning *"caused student isolation, demotivation, and resentment."* The answer is in delivering an interactive experience that incorporates a facilitator who gives the student the choice of managing the degree and depth of presentation, discussion, and interaction.

Randy Pausch achieved a memorable outcome through the combination of content, delivery, and passion. To be effective, learning must be engaging. In July 2008, a web-based education tool piloted with Aboriginal school children in northern Australia and using an educational games format including competitions, awards, and surprises achieved measurable improvements in vocabulary, comprehension, and visual skills while building competence and confidence in the use of computer technology.

My daughter told me about her favorite geology lecturer. He presented the sometimes mundane material in an entertaining and easy-to-understand way using multimedia applications such as video and simulated graphics, showcasing personal pictures of notable geographical sites taken during his travels, regularly asking questions of his audience to test that they understood the material being presented, and activating them by passing around geological samples so they could touch and feel the topic of the day. This lecturer also had a Facebook profile and connected with most of his students both academically and socially, giving his students the choice of nurturing relationships well after the university years.

Real-World Situations. Once relevance and the most effective delivery methodology are established, the applicability to real-world situations is the clincher. Connected learning should have one eye on effectiveness and another on how the learning will be used in the real world. It may mean modifying existing practices as well as articulating the value of the chosen learning process. An example is in examination conditions where students are sometimes not permitted to bring calculators and mobile phones into the room. A school in Sydney, Australia, piloted the approved use of connection technology during exams in August 2008. A spokesperson for the school argues that *"in preparing them for the real world, we need to redefine our attitudes toward traditional ideas of cheating. In their working lives they will never need to carry enormous amounts of information around in their heads. What they will need to do is access information from all their sources quickly and they will need to check the reliability of their information."*

Another example is the case of the remotely located student who commenced his online master's of business administration course (MBA) and was concerned about how to effectively coordinate group projects and assignments with other students in different time zones. He learned something from the course that he didn't expect to. By learning how to collaborate online, he gained a valued real-world skill for his organization that regularly required this capability, as they conducted business with global organizations. This portable skill differentiates him favorably from other connected individuals.

As mentioned, there is no silver bullet for connected learning effectiveness. However, there are a set of options on how to best utilize the resources available through a combination of effective delivery and efficient connection technologies.

CHAPTER SEVENTEEN
CHALLENGES

Don't be evil.
– Google company motto, Sergey Brin and Larry Page

1. Accessibility

I visited a local Apple retail store with the intent of purchasing the recently launched iPhone as a birthday present for my wife. As I was waiting to be served, I wondered in excitement how this amazing multifunctional and ubiquitous technological marvel would open up her world of access and connections to people and information. I looked around me and saw others perusing the demonstration devices with awe at what the humble mobile device could now do. To kill some time, I approached a computer and looked through the vendor's web site that provided many additional applications including news, sports, maps, directories, and online network access. I had the whole world of web access and mobile devices in this store and, shortly, in my hands. At that moment, I marveled at how connected I truly was.

Once I was served, I was told that iPhones were out of stock and would not be available in time for my wife's birthday. I had promised her one as a gift and my connected world suddenly lost its frame of reference. I demanded to see the manager and angrily requested an immediate escalation of the situation to the highest level. They told me that they'd see what they could do and get back to me "shortly." With no other option but to leave in disgust, I decided to make a dramatic exit. I approached the glass door leading to the street and pulled the metal handle hard—really hard—in fact, clear out of its attachment to the door. The store staff tried to reattach the broken door handle but it detached itself clear off the glass. I couldn't get out and I was left in limbo as the staff frantically tried to open the doors some other way. A brief feeling of isolation engulfed me. No one else in the store could get out into the street for five minutes until they finally levered the glass door panels enough to pry them open.

Upon reflection, the irony of the situation dawned on me. Minutes before I'd felt total connection to the outside world through technology. But through an act of rage, I not only felt physical disconnection to the real world outside, I incapacitated others at the same time. This situation made me evaluate some of the challenges confronting the *Connection Generation*. At a time when we feel so connected to each other and to information through technology, how we manage humanity's propensity to deconstruct itself and to commit beyond what it's capable of handling, while still understanding our sense of place in the world, is as real for each of us as that broken door handle in my hand that day.

2. Antisocial Behavior

In the real world, history has provided many lessons on how to manage behavior that should not be tolerated by society and business. From simple actions such as the issuing of fines for not observing parking signs and speed limits to the more complex punishment of being executed for crimes against humanity, the world is full of regulations and laws that manage antisocial behavior. Due to the relative maturity and openness of the World Wide Web, the online world is still learning how to maintain its democratization while limiting its potential abuse. There are three main factors that need to be considered when reviewing the challenges of antisocial behavior: intended malice, fraudulent schemes, and societal ethics.

There are situations where people desire to take advantage of the fact that people are connected for malicious purposes. By virtue of being connected on the internet, individuals are potentially exposed to the particular technical hazards—sometimes minefields—that are created by some to specifically cause havoc. Most users have been affected by a computer virus at some point in their lives. In May 2007, Google surveyed 4.5 million web pages and found what they called "vulnerabilities" in 10 percent of them that were capable of deploying viruses without the knowledge of unassuming visitors. These so-called "drive-by download" viruses are embedded in web pages and downloaded directly to the user's computer without warning. These viruses have a similar impact as their predecessors, which include replication resulting in automatic distribution to online address books, attacks launched on other unsuspecting sites, and disablement of all aspects of a computer's operation. Google has established a system that issues warnings about these potentially harmful sites, clearly identifying them in search results.

These virus attacks may seem more of an inconvenience than a threat to most, unless they happen to run online businesses that rely on the smooth operation of the interactive web or if they have irreplaceable data stored on their computers. And it seems that anyone with access to a computer and a malicious intent can launch these attacks. The "Yes, I heard you" denial of service attack, launched by a Canadian teenager in February 2000, caused the cessation of operations of multibillion dollar corporations. When a fifteen-year-old boy is capable of launching such a devastating disruption that affects the day-to-day operations of global businesses that rely on the web for their livelihoods, it requires some level of protection and insurance to minimize the risk and damage of such activities.

Similarly, there's the unsavory behavior of predators incapable of controlling their invasive urges. In Australia, it's estimated that 87 percent of Australian children have access to the internet with most having connection technologies such as e-mail, a blog site, and a mobile phone by their third year of formal schooling. Australian psychologist Dr. Michael Carr-Gregg says, *"Around seventy-one percent of kids receive personal messages from unknown people, twenty-five percent (of which) have been sexually solicited over the net. Cyber-bullying is increasing and parents often have no idea what is going on."* This activity has prompted the Australian government to provide $40 million to the Federal Police to track internet predators and a further $90 million for software filters.

There are connected individuals who are unfamiliar and unaware of various fraudulent schemes that are ever present with connection technologies that are capable of increasing their vulnerability and risk of being directly impacted by them. The instinctive behavior of securing personal items such as a wallet, handbag, and keys due to the inherent capability to permit direct access of personal and physical property applies to online information that's subject to the same, if not a more disciplined level of security scrutiny. Just as we wouldn't want credit cards to fall into the wrong hands and keys to our personal and business premises to be made available to criminals, making these items available on the web creates a potential risk for the connected individual, if left unsecured.

Once a person is connected on the internet, pervasive activities such as using online banking, posting information on online networking sites, and transacting with online stores creates a trail of activity that leaves what I've referred to in a previous chapter as "Webcrumbs." These information crumbs leave a virtual trail that allows just about anyone to access the information that's left on the web. If information posted and sent between people is more than on a "need-to-know" basis and with an appropriate amount of security, this information could expose one to identity theft. This is where someone steals your online identity—by knowing personal information such as your name and date of birth—and assumes it as their own for the specific purpose of committing fraud. It may even be a case of knowing someone's e-mail address, possibly procured from a business card or web site so that organized syndicates can develop scams to steal your personal information and passwords, more commonly known as phishing. These scams can be as simple as a stranger sending what appears to be a personal e-mail while posing as a Nigerian bank or insurance company. It advises of a bogus entitlement that can be claimed only by a return e-mail with your personal and banking details for direct transfer. According to a 2007 Australian Bureau of Statistics report, 800,000 Australians experienced some form of personal fraud in a calendar year, with 500,000 victims of identity fraud resulting in losses of almost one billion dollars.

The internet, while encouraging connectivity that should be designed to unite society, also tests the ethics that potentially may divide and destroy. The debate about whether the whereabouts of convicted sexual predators should be publicly and easily accessible on the internet is a case in point. Some argue that it is a vitally important public service to protect the general population from known pedophiles while others argue that it's a violation of an individual's human rights. Another physical threat is that of borderless, open markets such the pervasive online availability of prescription drugs that are regulated in the offline world so that their administration is controlled. Customs and pharmaceutical authorities are powerless to deny their entry into a country if obtained legally, despite the potential to harm users if their contents are questionable or unknown. Other banned items bought online include contraband items that, if undetected, could cause biosecurity issues. This threat is being taken seriously by some governments exemplified by the Australian Quarantine and Inspection Service spending more than $600 million per year to improve monitoring of items entering the country due to the outbreak of Britain's foot-and-mouth disease in 2001. From an ethical perspective, there are reports that seem to appear regularly of people willing to sell human organs, their virginity, and even babies on sites such as eBay, that sparks debate and discussion about the line between what constitutes basic freedom and an open and free marketplace.

3. Connection Overload

In 2008, it is estimated that over three billion personal computers exist, almost all of which have access to the internet. Companies such as AMD are designing and deploying the $100 computer to support programs to provide one laptop per child by the year 2015. Computers are commodities. They are housed in public areas such as internet cafés and will be standard in cars and airplanes in the not-too-distant future. The computer and the mobile device are converging rapidly and becoming more pervasive as evidenced by the popularity of the iPhone, which, after just eighteen months, was the most popular mobile handset in the United States. The Japanese claim that they'll be able to top the "smart glass" interface of the iPhone with a device that's capable of incorporating the display of three-dimensional images on a screen that fits in your hand.

All of these devices have access to billions of web pages. According to International Data Corporation (IDC) estimates, collaboration applications will create almost 400 billion gigabytes of "knowledge" while Google has a goal of making every book ever written available online. Companies like Google and Microsoft also want to store your personal documents and records on a virtual system that you can access from anywhere, called "cloud computing." The collective information machine that is the internet will allow people to *hunt, interact, test/trial, and share* information with people, anytime, anywhere, instantly. It will allow groups to *gather, regulate, organize, unite, and participate* as they have never before. And they're going to be able to do this all even faster than they can today. Australian scientists claimed in 2008 that the internet will only get faster—one hundred times faster—with technology being tested and developed in their labs. In 2008, IDC also estimates that a new technology being developed by information technology and telecommunications companies called "Unified Communications," which they describe as *"a common infrastructure to deliver, manage and support a wide range of communications applications—all accessible through desktop and mobile devices,"* will gain traction in the next two years.

Is all this connectedness becoming too much for both people and the technology? It seems that cracks are appearing across the board. A Polish driver placed all his trust in his Global Positioning System (GPS) road navigation device and—despite three road signs warning him of a road closure—diligently obeyed its instruction to drive straight ahead and into a recently created lake that served as a water reservoir. In Italy in 2008, the disgruntled government, which had just lost an election, posted—and granted public access to— the 2005 tax returns of forty million Italians online. The treasury department was forced to shut the site down eventually, following a deluge of privacy complaints.

According to psychological research conducted in the United States, 76 percent of people admitted to suffering from what psychologists are now calling "dicomgoogolation." It denotes a feeling of extreme stress that's usually attributed to being fired, attempting an examination unprepared, or being late for an important meeting. It is also used to describe the stress of being unable to go online because the system is down or a laptop or mobile device has run out of battery life. All these factual and anecdotal factors are indicators pointing to a world that's faced with the challenge of better understanding how individuals and groups connect and how they can be managed for both their personal and collective well-being. They pose real threats to societal well-being and, from a business perspective, productivity loss and burnout. It could also lead to disconnection.

From 1992 to 2004, I worked with a marketing colleague who became a close friend. Throughout those years, we both experienced the thrill of the early days of "web" marketing. This was marketing that went beyond the traditional elements of advertising, direct response marketing, public relations, and events and incorporated the World Wide Web and mobile marketing, through site advertising, e-mail and text message marketing, e-catalogues, e-commerce, and webinars. It was exciting in that we were pioneers in the next frontier of marketing. It was also challenging in that we were tasked to understand it, plan it, sell our initiatives internally, and implement it—all while complementing existing organization structures and channels of distribution. Our roles required that we were connected to the key stakeholders of our business internally and externally. We liaised with customers, suppliers, business partners, executives, and employees. We also had to stay connected to the rapid developments in technology, from the computer to the web to the mobile—and beyond.

In 2004, we changed jobs and despite our busy roles and schedules, we stayed in touch via regular courtesy phone calls and occasional face-to-face lunch meetings. In 2007, I received a phone call I wasn't expecting. She told me that she and her husband were retiring from corporate life and were moving to a remote town to a significantly quieter pace of life. I immediately congratulated her on a great career achievement and asked for her phone number so we could remain in contact. To my total surprise, she told me that she was getting a private number, would not have a mobile or laptop device or even an e-mail account. She made a life decision after so many years of being a slave to other people and to technology to "virtually check out." She was clear about not wanting to stay in contact and didn't supply the private number. To this day, despite my numerous online network connections and web savviness, we remain disconnected. I wonder how many others will do the same as it all gets too much in the future.

4. Globalization and Sense of Place

For my disconnected colleague, it was her choice to walk away from the connected rat race. But what of connected individuals who aren't in a position to choose? How do they navigate their sense of place in a connected world? What will they have to deal with?

The first key factor for the connected individual is the balance between personal identity and public perception. In a world that will allow them—and others—to post personal and professional information online, there are public perceptions that can develop as a result of what is assumed to be private information due to its easy and instant accessibility. The postings of Hollywood celebrities Paris Hilton (private sex video), Alec Baldwin (private voice mail message to his daughter), and David Hasselhoff (YouTube video filmed of him in a drunken state by his daughter) are all classic examples of reputations made online.

There's the more innocent case of Australian Olympic swimming gold medalist Stephanie Rice. Rice was a young athlete whose star was rapidly rising in the swimming ranks due to her record-breaking achievements in the pool during the Olympic trials in 2007. As public attention was focused on her due to her stellar performances, the simple actions that would be common to anyone her age suddenly painted a public persona that she didn't intend to reflect. In the same year, Rice had attended a costume party dressed as a sexy police officer. She posted three photos of herself from the party on her Facebook profile, which were accessed and prominently published in major newspapers and magazines across Australia. Rice was immediately projected by the media as a party girl and rumors of her dedication to win Olympic gold medals in Beijing in 2008 began to surface. As it happened, she put paid to the rumors by winning three Olympic gold medals. Had she not performed to expectations, a set of photos presumed available to her selected friends could have ruined the reputation of an aspiring superstar athlete.

On the other hand, there's the case of Joshua Lipton. Lipton faced a charge of drunk driving that seriously injured a woman. Two weeks after the event, he attended a Halloween party dressed as a prisoner. Someone posted the incriminating photo on Facebook with the label "Jail Bird." The prosecutor in the case downloaded the photo and used it as evidence to influence the judge's decision to sentence Lipton to two years jail. Kevin Bristow, Lipton's attorney, offered this advice to his own teenage children: *"If it shows up under your name you own it and you better understand that people look for that stuff."* Both are classic examples of the "Webcrumbs" factor.

The other challenge for the *Connection Generation* in managing their sense of place is the advent of globalization. People and business are so connected that it remains a challenge to understand where one's sense of place in the world is, socially and economically. The economic crisis of 2008 began in the United States but had repercussions in Europe,

India, and China. Bonds from banking institutions Fannie Mae and Freddie Mac contained one-fifth of China's currency reserves. *Time* magazine's Niall Ferguson observed that *"China and America have come so close to merging financially that we can almost speak of 'Chimerica.'"*

During the crisis, Dr. Marc Faber, an investment analyst and entrepreneur, commented on a proposed federal government rebate of $600 to every adult citizen to assist in alleviating the crisis in the United States. Possibly tongue-in-cheek, it was not too far from the truth: *"If we spend that money at Wal-Mart, the money goes to China. If we spend it on gasoline it goes to the Arabs. If we buy a computer it will go to India. If we purchase fruit and vegetables it will go to Mexico, Honduras and Guatemala. If we purchase a good car it will go to Germany. If we purchase useless crap it will go to Taiwan and none of it will help the American economy. The only way to keep that money here at home is to spend it on prostitutes and beer, since these are the only products still produced in the U.S."* Globalization continues to provide connected individuals and groups with many more challenges, some known and some yet to come, that will test humanity's sense of place in a connected world.

5. Management

In designing the values for their new company in 2004, Google founders Larry Page and Sergey Brin agreed on a key discipline to ensure that the power of their search innovation would be utilized for the greater good of the world and not for detrimental purposes. They stipulated that this ideal should be a cornerstone to how Google operated, and proclaimed this core value in a company motto called *"Don't Be Evil,"* which is embedded in employee induction training and ongoing education at Google. Unfortunately, we live in a world where such ideals are aspired to but not mandated or even practiced by some, which presents greater challenges to individuals and groups who seek to be more connected. To manage a generation of greater connectedness, it's important to comprehend the many inherent challenges faced by society and business. As in real life, individuals and groups in every society and business are all faced with challenges. To combine these challenges so as to make them appear insurmountable is self-defeating. To effectively address them is to break them down to what they really mean within the ecosystem of the *Connection Generation* and to manage them *inline* with we how we operate in both the physical and virtual worlds.

CHAPTER EIGHTEEN
SMART CONNECTION

We cannot live for ourselves alone. Our lives are connected by a thousand invisible threads, and along these sympathetic fibers, our actions run as causes and return to us as results.
– Herman Melville

1. *Inline*

We know we always have been, and are increasingly becoming more, connected. We know that some of us are more connected than others, there are methodologies and tools that can help us manage this connectivity for personal and commercial benefit, and that there will be challenges in a hyper connected world. The opportunities for the *Connection Generation* are plentiful—if applied and managed intelligently. Despite our obvious connection to technology, being "online" is not the be-all and end-all. We must embrace the richness of our backgrounds and cultural practices in which we were raised and leverage the learning, experiences, and instincts acquired from living in the "offline" world. To benefit from the opportunities of being connected as individuals and in groups, we need to take advantage of both our acquired online *and* offline assets to successfully manage and operate in an increasingly connected world. We need to utilize these "*inline*" with clearly defined goals, which should be based on *SMART* principles—*specific, measurable, achievable, relevant, and timely*. This ability to apply ourselves *inline* with what we want to achieve is what I call the *SMART* Connection.

2. *Scenario Management*

The story of an incident on a European family holiday two years ago is a case study that illustrates how to work through a challenge by managing and adapting in a connected world, using *SMART* connections. My wife, daughter, son, and I had been planning a holiday in Europe for the past four years. From the beginning, the *three Rs of connection* came into play in planning the trip. As the kids were approaching their twenties, the *reason* for the trip was to experience—as a family—the trip of a lifetime to Europe. In planning what we wanted to see and where we wanted to go, we spoke to people we had *relationships* with to draw on their previous experiences and expertise in the areas we intended to visit. The *return* for each of us would be to enjoy ourselves and experience it together as a family unit.

So when the time came to organize the itinerary, *selective* Kerry used her family, friends, colleagues, and travel industry connections to plan the trip. She then used her search skills to hunt down the best deals that would accommodate our plan—and budget—to see as much as we could in the available vacation time. It was planned using the knowledge of both our individual behaviors and family group dynamics. We factored in times when we had to get up early for sightseeing tours and, after particularly hectic travel days, set aside time to sleep in the following day. She booked almost all of the transport, accommodation, and tourist excursions online. She arranged an itinerary that had the four of us visiting fifteen cities in seven countries in twenty-five days. It was always going to be a tight itinerary, which in itself needed managing.

The first twelve days worked beautifully. We went to Paris, Bordeaux, Madrid, and Barcelona with all transfers running like clockwork. On day thirteen, we arrived in Nice, France, and spent two days there. Our departure was scheduled for day fifteen via a 9:00 a.m. train connection to Milan, Italy. This was an integral link in the trip that if missed would cause havoc to the remainder of our tight trip schedule. As our hotel was conveniently located across the road from the train station, I thought to walk across a few hours before and double-check that the train was still scheduled to leave on time. As I walked in, I noticed the lack of travelers at the station, which I dismissed as the off-peak travel season in France as it was the middle of December and also that no one was usually awake in most parts of Europe at nine in the morning. I did see a large notice board in front of the ticket office with the words *"grève des employée des chemins de fer."*

My brain registered that the French word "greve" and the similar sounding English word "grave" meant that something wasn't right. "Et meens, how you say in English, a train strike, Sir—no trains today," said the information desk attendant. With a limited French vocabulary based on a year of language study in high school, in my best "Franglais," I inquired if there was replacement transport organized for people to get to the major cities. His curt reply was "non."

In a clearly annoyed tone, I asked how my family would make a travel connection to Milan so we could keep our tight schedule on track. "Peut-être, demain," was his response. Even my limited high school French recognized what he said: "Maybe, tomorrow."

As I quickly approached a degree of anxiety, I advised the person of the importance of getting to Milan today as it would have a detrimental effect on the rest of our travel arrangements. Although he was clearly disinterested—obvious by the minimal eye contact and a shrug of the shoulders—I desperately asked if he could assist me in how I could make the Milan connection by another mode of transport. Without uttering another word, he raised his arm and pointed to the tourist office on the other side of the train station. I rushed over to the office to find a single office attendant behind a desk.

As I walked back to the hotel to work out what to do next, the thought entered my mind that even in an efficient on-line world where you could make travel bookings months in advance over the web from anywhere in the world that connects you to multiple cities at any time, the realities of the physical world—like train strikes—needed to be a real planning consideration. The tourist officer was more helpful. I told her that we had to be in Milan that evening. She began to present me with charter plane, helicopter, and hire car options, which were all out of our price range. I asked if there was an indirect connection to Milan from Nice that might get me there in two steps.

She searched for more options online. She then advised that the train strike was for French trains only and that the Italian trains were running. Her suggestion was to hire a taxi to take us across the border to the Italian town of Ventimiglia, which was an hour cab ride, as there was a train connection from there to Milan at 11:00 a.m. Good option, I thought, except it was so early in the morning that there were no cabs available outside. She told me that her brother's cousin was a local taxi driver and that he could probably take us for a reasonable price. In one phone call, she arranged for the cab to pick us up from the hotel at 9:00 a.m.

I walked back to the hotel thinking about how I'd inform the family of the situation. I instinctively went into connected individual PLAN management mode. To get their support to deal with the situation, I personalized the communication outlining the predicament we were faced with and the proposed solution. I listened to their comments, complaints, and suggestions and calmly advised them that they were all valid but that we now needed to take an appropriate action. I activated them by getting them to organize the packing of bags and hotel checkout as we needed to move quickly to prepare ourselves for the taxi. Once we were ready, I needed to manage tempers and arguments by using nurturing techniques such as positioning this as a learning experience that would make each of them more travel savvy in the future.

The taxi arrived as arranged and on time. After 150 euros in cab fare and in forty-five minutes, we had crossed three countries—France, Monaco, and Italy—arriving in Ventimiglia, Italy, with an hour to spare. We met the Eurail connection to Milan arriving two hours before the train from Nice would have gotten us there, which we utilized to see a little more of the city than we had planned. Many people have experienced situations such as this. This experience taught me about managing connections in both the real and virtual worlds and, more importantly, how they needed to be handled intelligently, or as I call it, "*SMART*."

3. Specific

In planning the trip, we were very *specific* about what we wanted to achieve on many levels. We knew where and why we wanted to go, how it would affect our relationships, and what we would get out of the experience (*three Rs*). Even at 108 years of age, *selective* centenarians Olive Riley and Ruth Hamilton knew exactly what they wanted to get out of their blogging activity in giving them the capability to share with the world the wealth of knowledge accumulated over all those years. Our trip also established a detailed connection map of the network of towns we were to visit in the form of an itinerary that planned and prearranged every transport connection by day and hour and what we would do in each place we went to, to better meet our objectives. Similarly, the CAN, FAN, and PAN planning method could be used to sort contacts to manage them into a network of individual connections.

The trip plan presumed a specific travel connection from Nice to Milan was established and locked in. When this was challenged, we needed to develop another plan specific to achieving our goal of "re-establishing" that critical link. In doing so, I needed to assess this particular situation and isolate the connections I had available that would provide the reconnection. The web was not an applicable connection technology in this circumstance, as I needed precise local knowledge to attain a specific solution. I hunted down the immediate connection network available to me, which in this case consisted of an information attendant and tourist officer.

Selective Bill Gates the technologist had more luck in choosing to utilize the wealth of business-savvy professional knowledge on the LinkedIn network numbering twenty-five million people to harness suggestions to his pressing business issue. In my case, I knew that the *basic* information attendant couldn't help me as we struggled to connect with a language barrier and an attitudinal disinterest in my dilemma. I was more likely to get a specific plan of action by connecting with the *selective* tourist officer who was more amenable to assisting me, which she utilized in activating her connections to both information (Ventimiglia train schedule option) and people (her brother's cousin, the taxi driver) to assist in making my personal connection to Milan. It's in knowing how and when to make these connections functional for their precise purpose that presents specific solution options to address both challenges and opportunities in a connected world.

4. Measurable

Our trip plan was timed across every connection step to suit our specific goals in a measured way. We knew the day and hour in which we needed to be when and where for the entire twenty-five- day trip so that the trip would meet our expectations. *Connected consumer* Kira uses her connection devices such as her iPhone as an alarm clock to remember when to wake up in the real world, her electronic diary to keep her work day on schedule, and is prompted by text message reminders of pending appointments. Her life relies on time as a measure of what needs to be achieved in small chunks as part of her daily life. Individual connections are also manageable in a measurable way.

Connected employee Luke manages his customer connections in varying ways to suit their circumstances so he can meet his measurable targets of short-term sales budget attainment and longer-term customer satisfaction results. Our trip plan included critical "deadlines," which, if not met, would take us off the path to getting us to where we wanted to go.

Connected leader Barack had an entire campaign's worth of deadlines, including winning primaries, campaign funding targets at individual and group levels, and the ultimate deadline of having his constituents bought in to his message enough to secure their votes on presidential election day. Our goal was to remain on track with the plan and to deal with the challenges and issues as they arose so we could relish the experience of visiting fifteen cities on our trip of a lifetime.

Connected business champion Jen was nimble enough to transform her business, while realizing the measurable business benefits of operating a more mobile operation. This would present her with more opportunities to stay in touch with her customers and her target market and to achieve her business revenue and expense plans.

5. *Achievable*

Although a tight plan by design, every travel connection was achievable. It was based on a published schedule, although dependent on external factors such as the weather and industrial disputes that were beyond our control. *Selective* Richard the sales manager prioritises his business and personal connectivity to achieve the desired outcome of a better work and social life balance. Just as the trip schedule took into consideration individual circumstances, Richard found a way to better manage how he dealt with his goals through his specific connections.

Similarly, *passive* Diana is a time-poor senior executive who understands her personal connection behavior and chooses to address her connection activity by "outsourcing" it to a personal assistant who has the capacity and the skills to manage it for her. It is in this way that she can achieve her connectivity challenge. The trip planning also took into account the dynamic of how we functioned collectively and the importance of everyone playing his or her part to achieve the desired outcome. In the case of the tsunami relief initiative at IBM, it was this understanding that led people to utilize their connectedness to personally and collectively contribute for the specific cause and the greater good, which led to achieving a successful donation campaign. The challenge presented during the trip of making the Nice to Milan connection so as not to miss other critical connections was one that needed to be achieved within a time limitation of three hours. In considering the options available, the plane, boat, and hire car modes of transport were not feasible for both financial and practical reasons in achieving the desired result of getting us to Ventimiglia.

Like life itself, the obvious options are not always the best ones given all factors. The utilization of the available connections such as the *selective* tourist officer led to an achievable solution. It is in expanding one's available network of individuals, groups, and networks that opportunities to address challenges become more achievable in practical reality.

6. Relevant

We made all the bookings online ahead of time as this was the most efficient way to manage it and we checked that these arrangements coincided with the physical world. It was relevant for our goals to preplan and monitor our connections along the way to ensure we met our objectives. *Basic* Geoff the taxi driver understands the business opportunity of adapting connection technology to work the way his customers want to be serviced. He realizes that these technologies are relevant both to his customers and to the efficient operation of his business.

For *active* Michelle the working mom, she had to experience the power of these technologies firsthand to be in a position to apply these practically to her day-to-day work and life activities in a way that made the most sense to her. Having tested and trialed these options, she positioned herself to become a hunter of these practical connection technologies and directly relate them to how she operates.

For *active* Hamish & Andy, they relate to their masses of fans using connection technologies that are adopted by their target audience. They realize that maintaining these connections is critical to their longevity and appeal and they must strive to keep abreast of newer technologies to stay relevant to their audiences.

7. Timely

I have detailed the importance of achieving deadlines to our trip goals. Another critical consideration is using connection resources to deal with issues that developed in a timely fashion. The Nice to Milan dilemma posed a threat to other connection linkages, potentially disrupting the connection map plan altogether if not addressed fast. We needed to take swift action and use all the resources available to divert an irrelevant connection option (train from Nice) into a more appropriate one (taxi to Ventimiglia) that eventuated in ultimately reaching our connection target (train to Milan).

Passive Pooja faced the dilemma of having to establish a personal brand in a foreign country in a short space of time to position herself for the job she desired in that market. She promptly utilized LinkedIn as the connection vehicle to build the profile that would be developed promptly into her public persona that would connect her to the right people at the right time. *Connected learner* Randy Pausch had a race against the life clock –due to his cancer– to leave the legacy of his "last lecture" by using a medium that spread his message throughout the world.

8. Conclusion

It seems that being more connected in the age of computers, the internet, and social networks is relevant only to the generations who grew up with it. There is no doubt that it will stand them well in dealing and coping in the next few years of the *Connection Generation*. So is it relevant that older generations become more familiar with connection technologies to better manage in a connected world? The research says yes. In 2008, researchers from the University of California in Los Angeles (UCLA) in the United States found that web-savvy adults aged fifty-five to seventy-six registered a twofold increase in brain activation when compared with those who were not. *"Our most striking find was that internet searching appears to engage a greater extent of neural circuitry that is not activated during reading—but only in those with prior internet experience,"* said Dr. Gary Small, a professor at UCLA's Semel Institute for Neuroscience and Human Behavior. From an anecdotal and practical perspective, it's provided tangible benefits to people like Olive Riley, Ruth Hamilton, and my father, Felix.

This book discussed the fact that we're all connected, regardless of generational age, in some way to other people, information, ideas, and experiences. This connection occurred through time and developed into an instinct necessary for the survival of society and business. It's revealed that some people are more connected than others and even connected groups and networks have their own distinctive ways of operating.

We know that better understanding these connections can help us to manage them for personal and commercial benefit. We can achieve this by implementing a *PLAN* methodology for individuals and applying *OPEN* principles with groups. We also know that this hyper connected world presents us all with more challenges, that if managed in a *SMART* way and *inline* with what we want to achieve individually and collectively, gives us a greater ability to thrive, revive, and survive in a connected world as part of the *Connection Generation.*

ACKNOWLEDGEMENTS:

It is impossible to acknowledge every single family member, friend, colleague, and connection who inspired me to write this book. Some are mentioned in the book and some are reflected as composite characters to maintain their anonymity. To all of you, I extend my deepest gratitude for helping me tell the story.

I owe the world of gratitude to my wife, soul mate, and partner, Kerry, who has been my daily inspiration since we met in 1984; my daughter, Rachel, and my son, Andrew, who have kept me youthful and whose patience I tested many times when all I ever talked about was the book. I also thank Rachel's boyfriend, Sam, for having to put up with the discussions and debates we had over dinner.

I thank my parents, Felix and Mila, and my mother-in-law, Helen, for always being there for me. I sincerely appreciate the support of my seven brothers and sisters, Mary Rose, Felix Jr., Milly, Tony, Jay, Victoria, and Santi for being my support base and to my brothers and sisters-in-law, Tracy, Louise, Wendy, Ian, Cathy, Di, Peter, Denise, Amanda, and Elena for the part they played. I also thank my grandmother, Mia, for being my guiding light. I also acknowledge the contributions of all my nieces and nephews who are too numerous to name but who know who they are. I also thank Maria Yoldi, Montse Sanchez, and James Massola for their contributions.

I am sincerely grateful to my assistant and friend Denise Derley for encouraging me to pursue the dream of putting my thoughts and experiences down on paper. I specifically acknowledge the direct contributions of Jaqui Lane, Peter Harris, Aja Shanahan, David Williams, Joe Talcott, Jen Harwood, Michelle Zamora, Pooja Kumar, Luke MacFarlane, Stan Relihan, Michael, Beth, and Jayde Ellenby, Jason Cupitt, Katie McMurray and Trevor Young.

I also acknowledge those who assisted me in their own way, namely Roger James, Mark Crowe, Malcolm Auld, John Butel, Melissa Richardson, Ian Lyons, Jennifer Taylor, Keith Stanley, and Ralph Kerle. I thank all my past colleagues from IBM and Telstra for providing me with the experiences and insights that formed the basis of the book. I thank my school mates from St. Leo's College for continuing to show me the value of staying connected. I also thank my new friends at Book-Surge, Zach Coddington and the teams responsible for making this publication possible. A special mention is due to my present and future connections on LinkedIn, Facebook, Twitter, Plaxo, MySpace, Xing, Ecademy, Nethooks, Gooruze, and all online networks whose interactions continue to provide me with a great source of intrigue, inspiration, and hope for a more connected world.

Finally, I thank our loyal family dog and companion, Piper, who spent many hours sitting patiently by my side as I researched and wrote the book.

QUICK REFERENCE:

This is a quick reference of the major concepts discussed in the book. It is recommended that if there are any terms used that require further clarification, please go to Google and type "define" in front of the word or phrase you are looking for.

Connection Generation or **Connect Gen** – anyone living since 1995 who had direct or indirect access to a connection device or relevant application, or to someone who had one.

Connection Technologies – online tools and applications that promote interaction with people, messages, and ideas through searching, sharing, collaborating, participating, sharing, and networking through user-generated content. Also known as social media, new media, Web 2.0, and Now media.

Online Networking – online platforms that enable the networking of individuals both socially through sites such as Facebook and MySpace and for business on sites such as Linkedin, Plaxo, Ecademy, and Xing.

ABCs of Connection – an acronym to describe the importance of ATTITUDE, BEHAVIOR, and CONTEXT in a connected world.

Three R's of Connection – factors at play in determining a person's attitude and behavior toward connecting. It incorporates REASON (why people connect), RELATIONSHIP (defines one's connection relative to their place in the world), and RETURN (what a person gets out of connecting).

HITS – an acronym to describe the online behavioural profile of connected individuals. It incorporates HUNT, (an extension of browsing behavior), INTERACT (beyond passive viewing), TEST/TRIAL (ability to try and test new tools and applications), and SHARE (passing information and ideas to their friends, contacts, and connections).

PLAN – an acronym to describe the key steps to take to managing connected individuals to best realize their personal and commercial roles in a connected world. It comprises PERSONALIZE (recognising the importance of personally connecting), LISTEN (understanding how to best connect), ACTIVATE (establish specific connections in the most appropriate way), and NURTURE (managing connections for longer-term potential benefits).

GROUP – an acronym to describe the behavioral profiles of connected groups. It comprises GATHER (congregate with ease and efficiency), REGULATE (establish rules and modes of group operation), ORGANIZE (form the group as facilitated by a connection technology), UNITE (under the common interest of the group within context and relevance), and PARTICIPATION (interactive individual and group communication).

OPEN – an acronym used to describe the key steps to managing connected groups to best leverage their personal and commercial roles in a connected world. It comprises OPPORTUNITY (to realize the goal and collective purpose of group), PARTICIPATION (establishing disciplines that promote rationale of being part of a greater group to achieve objectives), ENGAGEMENT (to realize the fullness of opportunities), and NETWORK (to realize the power of connecting within and with other groups to achieve outcomes).

Connections Profiles: the five distinctive profiles that defines an individual's degree of connectedness.

- ***Basic Connector*** – The *basic* connector has the least number of visible connections, has a CLASSICAL mental image of connection, is a laggard when it comes to technology adoption, and usually delegates connection needs to a known person.

- ***Passive Connector*** – The *passive* connector has a moderate number of connections, has a CONCEPTUAL view of connection, is in the late majority when it comes to technology, and exudes "passenger" behavior when it comes to connecting.

- ***Selective Connector*** – The *selective* connector has a manageable number of connections, has a RELATIONAL view of connectivity, is in the early majority of technology adoption, and is engaged in his/her connection behavior.

- **Active Connector** – The *active* connector has a large number of connections, has a PHYSICAL mental picture of connection, is in the early adopter category of technology, and drives his/her connection activity.

- **Super Connector** – the *super* connector is hyper connected, has a STRUCTURAL view of connectivity, is an innovator with technology use, and possesses a disciplined management style toward connectedness.

SCAN – an acronym to describe the set of factors that defines a connected business. It comprises SMART (specific, measurable, achievable, relevant, and timely business goals), CUSTOMER FOCUS (accountable understanding of the customer opportunity), ACCESSIBLE (the need to be where the customer needs them to be), and NIMBLE (ability to change and adapt to market and technology environments).

Four P's of Online Networking – the key factors in successfully participating and obtaining value from online networking. It incorporates PURPOSE (what does one want to achieve from joining an online network), PROFILE (creating an online persona that promotes your personal brand), PARTICIPATION (being actively involved in the online community), and PERSISTENCE (the discipline to persist with participation to get the most value from online networking).

Inline – to establish the most relevant and effective connection methodology, whether offline or online, "in line" with what needs to be achieved.

SMART Connection – an acronym to describe SMART objectives as they relate to a connected world (see Chapter 18).

END NOTES:

These notes are by no means exhaustive and may make indirect references to the source material in an attempt to acknowledge the original reference point where applicable. Web site page references are current at the time of publication.

Introduction:

1. Reference: Olive Riley's Blog: http://www.allaboutolive.com.au/
2. Definition: Generation (source: Wikipedia)
3. Reference: Six Degrees of Separation by John Guare (source: Wikipedia)
4. Reference: The Tipping Point by Malcolm Gladwell

Chapter One:

1. Quote: George Bernard Shaw: http://thinkexist.com/quotation/life_isn-t_about_finding_yourself-life_is_about/8906.html

2. References: Linked by Albert Laszlo Barabasi
3. Reference: Report on Microsoft IM Research: http://news.cnet.com/8301-13505_3-10005553-16.html
4. Reference: Report on 2007 Danish study of blue-eyed people relativity: http://www.foxnews.com/story/0,2933,327070,00.html
5. Reference: The Family Obama, Time magazine, 2008.
6. Reference: Research: Eskimos (source: Wikipedia)
7. Reference: Research: Jewish Law (source: Wikipedia)
8. Definition: Generation from American Heritage Dictionary http://www.bartleby.com/61/2/G0080200.html

Chapter Two:

1. Quote: Marshall McLuhan: http://www.livinginternet.com/i/ii_mcluhan.htm
2. Reference: Johannes Gutenberg, Anne Frank, Alexander Graham Bell, Gugliemo Marconi, and Orson Welles (source: Wikipedia)
3. Reference: Internet History (source: Wikipedia)
4. Reference: International Telecommunications Press Release: http://www.itu.int/newsroom/press_releases/2008/29.html
5. Reference: The Tipping Point by Malcolm Gladwell
6. Reference: Wikipedia and YouTube statistics sourced from Time magazine 2006 Person of the Year edition, December 26, 2006 to January 1, 2007
7. Reference: Video Snacking—Time magazine, January 21, 2008
8. Reference: Apple iPhone 3G Sales – http://www.techcrunch.com/2008/07/14/the-mobile-web-is-here-apple-sells-one-million-3g-iphones-first-weekend-ten-million-iphone-apps-downloaded/

Chapter Three:

1. Quote: Leonard Kleinrock: http://www.livinginternet. com/w/wi_online.htm

2. Reference: Market Penetration Rates reference from: What I Didn't Learn from Google but Wish I Had by Jamie McIntyre

3. Reference: Worldwide Internet Stats: http://www. internetworldstats.com/top20.htm

4. Reference: Google 40 Languages: http://googleblog. blogspot.com/2008/07/hitting-40-languages.html

5. Reference: Google Aotearoa Launch: http://www. bizreport.com/2008/07/google_aotearoa_launched. html

6. Reference: Internet Comes to Your Car: http://www. australianit.news.com.au/story/0,24897,23980118-15321,00.html

7. Reference: Kevin Kelley—Next 5,000 days on the Web: http://www.ted.com/talks/kevin_kelly_on_the_next_5_000_days_of_the_web.html

8. Reference: BlogHer and Compass Partners Study: http://www.bespacific.com/mt/archives/018304.html

9. Reference: Aging Drives Internet Growth: http://news. zdnet.co.uk/internet/0,1000000097,39116559,00.htm

10. Reference: Top 50 Great eBusinesses and the Minds behind Them by Emily Ross and Angus Holland.

11. Reference: How Online Shoppers are Bagging Bargains, The Sun-Herald, June 15, 2008, pp: 28-29

12. Reference: The Rich Shop Online, Google Study Finds: http://industry.bnet.com/retail/1000268/the-rich-shop-online-google-study-finds/

13. Reference: British Website reveals Australia's history: http://www.abc.net.au/news/stories/2008/06/04/2264456. htm

14. Reference: Facebook changes luck for lonely cat, The Sun-Herald, June 22, 2008, p.22.
15. Reference: Online Sweethearts Lived on the Same Street: http://www.abc.net.au/news/stories/2008/10/09/2385975.htm
16. Reference: Sergey Brin starts blog, tells of Parkinson's risk: http://news.cnet.com/8301-1023_3-10045958-93.html
17. Reference: State of the Blogosphere: http://technorati.com/blogging/state-of-the-blogosphere/

Chapter Four:

1. Quote: Martin Luther King: http://www.quotedb.com/quotes/2644
2. Reference: The Tipping Point by Malcolm Gladwell
3. Quote: Computer and Drug Users: http://www.quotationspage.com/quotes/Clifford_Stoll/
4. Reference: Technology Adoption Cycles (source: Wikipedia)
5. Reference: Colmar Brunton Research study: Connections 2008.

Chapter Five:

1. Page 32: Heraclitus Quote: http://thinkexist.com/quotation/a-hidden-connection-is-stronger-than-an-obvious/348531.html
2. Page 35: Life's Like That, Australian Reader's Digest, June 2008, p.71

Chapter Six:

1. Quote: Jean Armour Polly: http://www.livinginternet.com/w/wu_surf.htm

2. Reference: Martha Stewart and Marthapedia: http://mediatrope.blogspot.com/2007/10/marthapedia.html

Chapter Seven:

1. Quote: Bill Shafer: http://worldsoldestblogger.blogspot.com/2008/08/ruth-hamilton-109-was-worlds-oldest.html
2. Reference: Check LinkedIn Answers for Bill Gates: www.linkedin.com

Chapter Eight:

1. Quote: Sandra Day O'Connor: http://www.brainyquote.com/quotes/quotes/s/sandradayo372198.html
2. Reference: Hamish & Andy stats from their web site: http://www.2dayfm.com.au/shows/hamishandandy/
3. Reference: Oprah Winfrey by Michelle Obama—Time magazine, 100 Most Influential People edition, May 12, 2008, p.40.
4. Reference: Beginners Guide to Social Networking, Working Mums blog by Michelle Zamora: http://wondermum.blog.com/social percent20networks percent20michelle percent20zamora/

Chapter Nine:

1. Quote: George J. Sidel: http://thinkexist.com/quotes/george_j._seidel/
2. Reference: Pew Internet & American Life Project Study, April/May 2008: http://www.australianit.news.com.au/story/0,,23871412-5013040,00.html?from=pu
3. Reference: Obama Web 2.0 stats from: http://www.techpresident.com/

4. Reference: Obama LinkedIn stats from: www.toplinked. com

5. Reference: Gallup Poll 2008 from: http://www.abc.net. au/news/stories/2008/10/22/2398181.htm?WT.mc_ id=newsmail

6. Ron Bates LinkedIn Profile information: www.linkedin. com

7. Reference: Stan Relihan Connections Show: http:// connections.thepodcastnetwork.com/

8. Reference: Corey Worthington (source: Wikipedia)

Chapter Ten:

1. Quote: Mark Zuckerberg quote: http://www.facebook. com/press/releases.php?p=48242

2. Reference: Audience or Mexican Wave (source: Wikipedia)

3. Reference: Facebook statistics: www.facebook.com

4. Reference: World Youth Day social network: www.xt3. com

5. Reference: Anglicans out to connect, The Sun-Herald, October 12, 2008, p.35

6. Reference: Olympics from www.olympics.com

7. Reference: Al Qaeda and the Internet: The Danger of Cyber Planning by Timothy L. Taylor: http://www.iwar. org.uk/cyberterror/resources/cyberplanning/al-qaeda. htm

8. Reference: Justin Timberlake Facebook Fan Page: www. facebook.com

Chapter Eleven:

1. Quote: Craig Newmark re: Mark Zuckerberg—Time magazine, 100 Most Influential People edition, May 12, 2008, p.63.
2. Reference: Affluence.org: www.affluence.org
3. Reference: A-Space – Facebook for Spies, Time magazine, September 2008
4. Reference: Synovate Study: http://www.arabianbusiness.com/532785-uae-second-in-world-list-for-online-social-networking
5. Reference: Pizza Hut on Facebook: http://www.pizzahut.com/newsroom/2008/onlineOrderingOnFacebook.aspx
6. Reference: Christian Mayaud appearance on Connections Show: http://connections.thepodcastnetwork.com/2007/09/17/connections-002-conversation-with-a-lion/

Chapter Twelve:

1. Quote: Harriet Goldhor Lerner: http://www.wisdomquotes.com/003118.html
2. Reference: storemob: http://storemob.com/
3. Reference: Text Messaging in U.S. Politics, Newsweek: http://www.newsweek.com/id/46675?tid=relatedcl
4. Reference: MySpace (source: Wikipedia)
5. Reference: First Internet Wedding: http://www.quantumenterprises.co.uk/internet_wedding/

Chapter Thirteen:

1. Quote: Razorfish: http://www.digitaldesignblog.com/2008/10/28/announcing-feed-the-razorfish-consumer-experience-report/

Chapter Fourteen:

1. Quote: Michel Hogan: http://www.smartcompany.com. au/Blog/Michel-Hogan/20080804-B2B-brand-matters. html
2. Reference: Jen Harwood: www.jenharwood.com
3. Reference: How to Get the Job U Want: www. getthejobuwant.com
4. Reference: The Art of Networking: http://www. artofnetworking.com/
5. Reference: Wal-Mart's Ahead of the Pack on Social Media: http://industry.bnet.com/retail/1000267/groundswell-author-wal-marts-ahead-of-the-pack-on-social-media/
6. Reference: IDC Asia Report in Home Sweet Office: Telecommute Good for Business, Employees and Planet, by Brendan Koerner, Sept. 22, 2008: http://www.wired. com/culture/culturereviews/magazine/16-10/st_essay

Chapter Fifteen:

1. Quote: Malcolm Gladwell from The Tipping Point by Malcolm Gladwell
2. Reference: The Off-Line American, by Lev Grossman, Time magazine, August 25, 2008.
3. Reference: Why Dems Rule the Web, by Michael Scherer and Jay Newton Small, Time magazine, April 28, 2008
4. Reference: How He Did It, by Karen Tumulty, Time magazine, June 16, 2008.
5. Reference: The Democrats: Five Faces of Barack Obama, Time magazine, September 1, 2008
6. Reference: How Obama Really Did It by David Talbot, Technology Review: https://www.technologyreview. com/web/21222/

7. Reference: Obama Surfs the Web to the White House by Chris Lefkow, November 6, 2008: http://www.australianit.news.com.au/story/0,24897,24610417-15306,00.html?referrer=e-mail

8. Reference: Obama hits the web to fight the smears:http://www.abc.net.au/news/stories/2008/06/13/2273483.htm

9. Reference: Obama 2.0—the iPhone years: http://www.itwire.com/content/view/20985/53/

10. Reference: Obama does a 360: http://www.itwire.com/content/view/21171/532/

11. Reference: Obama "Neighbor-to-Neighbor" vs. McCain "Voter-to-Voter": Not a Fair Fight http://www.techpresident.com/blog/entry/31846/obama_neighbor_to_neighbor_vs_mccain_voter_to_voter_not_a_fair_fight

12. Reference: Obama in Orbit by Roger Cohen, November 2007: http://www.nytimes.com/2007/11/15/opinion/15cohen.html

13. Reference: Obama Victory Speech extract from Sydney Morning Herald, November 6, 2008

Chapter Sixteen:

1. Quote: Jean-Francois-Abramatic: http://www.livinginternet.com/w/wi_w3c.htm

2. Reference: iTunes Uni Hits Australia: http://www.australianit.news.com.au/story/0,,23808418-5013040,00.html?from=pu

3. Reference: Online courses better than being there: http://www.australianit.news.com.au/story/0,24897,24464426-15318,00.html?referrer=e-mail

4. Reference:BusinessschoolinallA380classes:http://www.australianit.news.com.au/story/0,24897,24351222-15306,00.html?referrer=e-mail
5. Reference:EducationRevolutionworriesteachers:http://www.australianit.news.com.au/story/0,,23889276-5013040,00.html?from=pu
6. Reference: Lecturers learn to sort out virtual classes: http://www.australianit.news.com.au/story/0,24897,24384751-24169,00.html?referrer=e-mail
7. Reference: Masters of the Universe by Jeanne-Vida Douglas, BRW magazine, Sept. 18-24
8. Reference: School to Allow iPods, internet in exams: http://news.ninemsn.com.au/article.aspx?cmp=nl_news_20august2008_14&mch=newsletter&id=617470
9. Reference:Web game aids in literacy teaching:http://www.australianit.news.com.au/story/0,24897,23984649-15318,00.html
10. Reference: Randy Pausch Last Lecture on www.youtube.com

Chapter Seventeen:

1. Quote: Google Company Motto: http://investor.google.com/conduct.html
2. Reference: Drive by Downloads: http://www.zdnet.com.au/news/security/soa/Google-warns-drive-by-downloads-up-300-percent-/0,130061744,339286098,00.htm
3. Reference:Warning over online medication purchases, The Sun-Herald, June 2008.
4. Reference: Your Weapons in the war against online fraud,The Sun-Herald, October 19, 2008
5. Reference: Net a threat to biosecurity,The Sun-Herald, January 13, 2008.

6. Reference: Dark side of a bright shiny world, Hills Shire Times, December 18, 2007.
7. Reference: IDC Australia Information and Communications Technology 2008 Top 10 Predictions Report.
8. Reference: Discomgoogolation: Internet's down and so are you, The Sun-Herald, September 7, 2008.
9. Reference: Driver follows GPS into lake: http://www.abc.net.au/news/stories/2008/10/25/2401130.htm
10. Reference: Stroke of Genius: Stephanie Rice, Sunday Life magazine, October 2008.
11. Reference: Facebook photos become evidence source: http://news.ninemsn.com.au/article.aspx?id=600723
12. Reference: The End of Prosperity? By Niall Ferguson, Time magazine, October 13 2008.
13. Reference: Publish and be taxed: http://www.economist.com/world/europe/displaystory.cfm?story_id=11332879
14. Reference: Last Word: Dr. Marc Faber, Fitz Files, The Sun-Herald, October 19, 2008.

Chapter Eighteen:

1. Quote: Herman Melville: http://www.wisdomquotes.com/001422.html
2. Reference: Googling good for geriatrics: http://www.smh.com.au/news/technology/googling-good-for-geriatrics-study/2008/10/16/1223750168833.html

INDEX: